PE MECHANICAL

THERMAL AND FLUID SYSTEMS

SIX-MINUTE PROBLEMS

WITH SOLUTIONS

FOURTH EDITION

DANIEL C. DECKLER, PhD, PE

PPI®

PPI2PASS.COM
A **KAPLAN** COMPANY

Report Errors for This Book

PPI is grateful to every reader who notifies us of a possible error. Your feedback allows us to improve the quality and accuracy of our products. Report errata at **ppi2pass.com**.

THERMAL AND FLUID SYSTEMS SIX-MINUTE PROBLEMS
Fourth Edition

Current release of this edition: 1

Release History

date	edition number	revision number	update
Aug 2022	4	1	New edition.

PPI
ppi2pass.com

ISBN: 978-1-59126-880-2

Table of Contents

About the Author

Daniel C. Deckler was born in Tiffin, Ohio, and raised in various parts of Ohio. He received his bachelor's degree in mechanical engineering from Ohio Northern University and received both his master's degree in mechanical engineering and his PhD in engineering from the University of Akron. He received his Ohio professional engineer's license in 1999.

Dr. Deckler is currently a professor of engineering at the University of Akron, where he teaches courses in mechanical engineering and advises the SAE formula combustion and electric race teams.

Before coming to the University of Akron, he was an engineer for Rockwell International in Downey, California; the Timken Company in Canton, Ohio; and the Loral Corporation (formerly Goodyear Aerospace) in Akron, Ohio. With these companies he worked on interesting projects such as the space shuttle and space station, bearing applications, and anti-submarine warfare systems, respectively.

Dr. Deckler lives in Akron, Ohio. When not working, he can be found enjoying the outdoors, be it golfing, cycling, running or training for triathlons, or hunting.

Preface and Acknowledgments

The Principles and Practice of Engineering examination (or PE exam for short) is prepared by the National Council of Examiners for Engineering and Surveying (NCEES). The PE exam tests examinees' understanding of both theoretical and practical engineering concepts. For mechanical engineering, three versions of the exam are given, one for each of the disciplines of HVAC and refrigeration, machine design and materials, and thermal and fluid systems. This book is designed to help you with the thermal and fluid systems exam.

For each exam, the NCEES issues a list of the knowledge areas that will be covered. The topics covered in *Thermal and Fluid Systems Six-Minute Problems* are the same knowledge areas as in the NCEES specifications for the thermal and fluid systems exam. The problems in this book cover fluid mechanics, heat transfer, mass balances, thermodynamic cycles, hydraulic and fluid equipment, energy and power equipment, cooling and heating, and more.

In 2020, the NCEES changed the PE Mechanical Engineering licensing exams from a pencil-and-paper format to a computer-based one. This in itself is not a big change. However, in the pencil-and-paper exam you were permitted to bring your own reference material, including *Mechanical Engineering Reference Manual* (*Reference Manual*), into the exam room. In the computer-based exam, you may not. Instead, you will have on-screen access to a searchable electronic copy of the *NCEES PE Mechanical Reference Handbook* (*NCEES Handbook*). This is the only reference you may consult during the exam.

This drastically changes how you must study for the exam. It is no longer enough to learn how to solve exam problems using the *Reference Manual* and other familiar reference books that you may annotate, highlight, and glue tabs to as you study. Now you must learn how to quickly find the equations and data you need in one specific source, an unmarked electronic copy of the *NCEES Handbook*. That is the reason for this new edition.

The fourth edition has been completely updated so the only reference you need is the *NCEES Handbook*. Problems include *NCEES Handbook* references to aid students in preparing to take the computer-based test (CBT) NCEES now uses. Also, the fourth edition includes 24 new problems. Eight of those problems replace those that no longer conform to the new NCEES testing topics; the other 16 cover new topics introduced by NCEES into the exam, provide additional examples for older topics, and bring the total number of problems in the book to 100. I hope these new problems will help you better prepare for the exam.

Some of the problems in this book originated in my professional practice, and others in my experience as an instructor. This book serves as a companion to the *Reference Manual* and is designed primarily to help you remember what you have already learned. I have purposefully left "shortcuts" out of the problem solutions. Instead, I have focused on solving the problems from fundamental concepts. Solving in this manner might take a few more minutes, but it eliminates the need to remember special cases. One only has to recall a few basic ideas that are easily remembered or located. In addition, for several of the problems I have shown multiple methods of solution, to further strengthen your review.

The reason for this book is to give you practice in solving exam-like problems within exam-like time limits. The problems are designed to be solvable in an average of six minutes each, which is the average amount of time you'll have for each problem on the actual exam. Some will take you less than that, of course, and others will take you longer. But overall, you should be striving for an average time of six minutes or less per problem.

The problems presented in this book are representative of the type and difficulty of those you will encounter on the PE exam. The book's problems are both conceptual and practical, and they are written to provide varying levels of difficulty. Though you probably won't encounter exam problems on exactly the same situations, reviewing these problems and solutions will increase your familiarity with the exam problems' form, content, and solution methods. This preparation will help you considerably during the exam.

Problems and solutions have been carefully prepared and reviewed to ensure that they are appropriate, understandable, and correctly solved. If you find errors or discover alternative, more efficient methods of solving a problem, please bring it to PPI's attention so that your suggestions can be incorporated into future editions. You can report errors and keep up with the changes made to this book, as well as changes to the exam, by going to PPI's website at **ppi2pass.com**.

This book would not have been possible without the contributions of the many people who supported me through several years of writing and review.

A huge thank-you goes to my engineering friends at a refinery near me. Many of the problems you see in this book are a result of their experiences. So, whenever you see a problem about a refinery, you will now know where it came from. Another thank-you goes to all my former engineering co-workers who were kind enough to take me "under their wings" and teach me what they knew when I was a beginning engineer. Believe it or not, they were an inspiration for several problems so many years later. Finally, I would like to thank my former and current students who are interested in real-life applications of the subjects I teach. They are always making me reach into my memory and pull out useful problems that I had to solve during my days in industry and as a consultant. Several of these problems have found their way into this book.

To David Bostain, Matt Gordon, and Muamba Wanzala, who reviewed and helped polish the manuscript, and Anil Acharya, calculation checker, thank you all. I very much appreciate your hard work.

I would also like to thank all the fine people at PPI, a Kaplan Company, who assisted throughout the publication process: editorial: Tyler Hayes, Scott Marley, Grace Wong; art, cover, and design: Tom Bergstrom; production and project management: Beth Christmas, Crystal Clifton, Jeri Jump, Stan Info Solutions, Kim Wimpsett; content and product: Nicole Evans, Anna Howland, Joe Konczynski, Maya Ma, Scott Rutherford, Megan Synnestvedt.

To all of you who read this book, it is my greatest hope that you find it useful. Good luck.

Daniel C. Deckler, PhD, PE

Introduction

EXAM FORMAT

The Principles and Practice of Engineering examination (PE exam) for mechanical engineering is an eight-hour exam containing 80 problems. After you see approximately half of the problems, you will have a chance to review and then submit them. Once submitted, you will no longer have access to those problems. You will then have the option of taking a scheduled 50-minute break before completing the remaining problems.

All problems are multiple-choice. Each problem includes a problem statement that contains all the information needed to find the answer to a question. The problem statement is then followed by four answer options, only one of which is correct. Problems are generally self-contained and independent, so an incorrect choice on one problem typically will not carry over to subsequent problems.

There are three versions of the mechanical engineering exam: thermal and fluid systems, HVAC and refrigeration, and machine design and materials. The mechanical thermal and fluid systems exam covers three major topics: principles (28–44 problems), hydraulic and fluid applications (21–33 problems), and energy/power system applications (21–33 problems). The knowledge areas and the approximate distribution of problems in the mechanical thermal and fluid systems exam are as follows.

Principles

- Basic engineering practice: terms, symbols, technical drawings, economics, units

- Fluid mechanics: fluid properties, compressible and incompressible flow

- Principles of heat transfer

- Principles of mass balance

- Thermodynamics: properties, cycles, energy balances, combustion

- Supportive knowledge: pipe system analysis, joints, psychrometrics, codes and standards

Hydraulic and fluid applications

- Hydraulic and fluid equipment: pumps and fans, compressors, pressure vessels, control valves, actuators, connections

- Distribution systems: pipe flow

Energy/power system applications

- Energy/power equipment: turbines, boilers, steam generators, internal combustion engines, heat exchangers, cooling towers, condensers

- Cooling/heating: capacity, loads, cycles

- Energy recovery: waste heat, storage

- Combined cycles: components, efficiency

For further information and tips on how to prepare for the mechanical engineering PE exam, consult the *Mechanical Engineering Reference Manual* or the PPI website, **ppi2pass.com**.

THIS BOOK'S ORGANIZATION

Thermal and Fluid Systems Six-Minute Problems is organized into three sections: Principles, Hydraulic and Fluid Applications, and Energy/Power System Applications. Each section contains problems pertaining to the knowledge areas within that division of the NCEES specifications.

Most of the problems in this book are quantitative, meaning that some calculation is necessary to arrive at the correct answer. A few problems are nonquantitative. Some problems will require a little more than six minutes to answer and others a little less. To complete 80 problems in eight hours (480 minutes), you should aim at spending an average of six minutes per problem.

Thermal and Fluid Systems Six-Minute Problems does not include problems related directly to HVAC and refrigeration, nor to machine design and materials. Other books in the *Six-Minute* series provide problems for review in these areas, and the *Mechanical Engineering Reference Manual* provides instruction and example problems in all three areas.

HOW TO USE THIS BOOK

The main purpose of *Thermal and Fluid Systems Six-Minute Problems* is to get you ready for the NCEES PE mechanical exam. Use it along with the other PPI PE mechanical study tools to assess, review, and practice until you pass your exam.

ASSESS

To pinpoint the subject areas where you need more study, use the diagnostic exams on the PPI Learning Hub (**ppi2pass.com**). How you perform on these diagnostic exams will tell you which topics you need to spend more time on and which you can review more lightly.

REVIEW

PPI offers a complete solution to help you prepare for exam day. Our mechanical engineering prep courses and *Reference Manual* offer a thorough review for the PE mechanical exams. *Thermal and Fluid Systems Six-Minute Problems* and the PPI Learning Hub quiz generator offer extensive practice in solving examlike problems.

PRACTICE

Learn to Use the *NCEES PE Mechanical Reference Handbook*

Download a PDF of the *NCEES Handbook* from the NCEES website. As you study, take the time to find out where important equations and tables are located in the *NCEES Handbook*. Although you could print out the *NCEES Handbook* and use it that way, it will be better for your preparations if you use it in PDF form on your computer. This is how you will be referring to it and searching in it during the actual exam.

A searchable electronic copy of the *NCEES Handbook* is the only reference you will be able to use during the exam, so it is critical that you get to know what it includes and how to find what you need efficiently. Even if you know how to find the equations and data you need more quickly in other references, take the time to search for them in *NCEES Handbook*. Get to know the terms and section titles used in the *NCEES Handbook* and use these as your search terms.

In this book, each equation from the *NCEES Handbook* is given in blue and annotated with the title of the section the equation is found in, also in blue. Whenever data are taken from a figure or table in the *NCEES Handbook*, the title of the figure or table is given in blue. Get to know these titles as you study; they will give you search terms you can use to quickly find the equations and data you need, saving valuable time during the exam.

Using steam tables, h_1 389.0 Btu/lbm, $s_1 = 1.567$ Btu/lbm-°R, and $p_2 = 4$ psia. h_2 represents the enthalpy for a turbine that is 100% efficient. Since the turbine is isentropic, $s_1 = s_2$. Using steam tables, find the appropriate enthalpy and entropy values at state 2' where 2' = 4 psia. [Properties of Saturated Water and Steam (Temperature) - I-P Units]

$$h_f = 120.87 \text{ Btu/lbm}$$

$$s_f = 0.2198 \text{ Btu/lbm}-°R$$

$$h_{fg} = 1006.4 \text{ Btu/lbm}$$

$$s_{fg} = 1.6424 \text{ Btu/lbm}-°R$$

The steam quality at the turbine exhaust (state 2) for a 100% efficient turbine is found from the entropy relationship.

Properties for Two-Phase (Vapor-Liquid) Systems

$$s = s_f + x_{fg}$$
$$x = \frac{s - s_f}{s_{fg}}$$
$$= \frac{1.567 \dfrac{\text{Btu}}{\text{lbm}-°R} - 0.2198 \dfrac{\text{Btu}}{\text{lbm}-°R}}{1.6424 \dfrac{\text{Btu}}{\text{lbm}-°R}}$$
$$= 0.82$$

The *NCEES Handbook* generally does not indicate whether an equation requires g_c when used with U.S. customary units. On the PE exam, then, you will need to know when and how to include g_c in a calculation without any help from the *NCEES Handbook*.

To show the correct use of g_c, equations in this book are given in two versions where appropriate, one for use with SI units and one for use with U.S. customary units, with g_c correctly included in the U.S. version. When you solve practice problems, however, you should use the *NCEES Handbook* as your only reference, identifying when and how to use g_c on your own. This is more trouble than looking up the equations in this book, but it will better prepare you for the actual exam.

Access the PPI Learning Hub

Although the *Thermal and Fluid Systems Six-Minute Problems* can be used on its own, it is designed to work with the PPI Learning Hub. At the PPI Learning Hub, you can access

- a personal study plan, keyed to your exam date, to help keep you on track

- diagnostic exams to help you identify the subject areas where you are strong and where you need more review

- a quiz generator containing hundreds of additional examlike problems that cover all knowledge areas on the PE mechanical exams

- two full-length NCEES-like, computer-based practice exams for each of the PE mechanical engineering disciplines, to familiarize you with the exam-day experience and let you hone your time management and test-taking skills

- electronic versions of *Thermal and Fluid Systems Six-Minute Problems*, *Mechanical Engineering HVAC and Refrigeration Practice Exam*, *Mechanical Engineering Machine Design* and *Materials Practice Exam*, *Mechanical Engineering Reference Manual*, and *Mechanical Engineering Practice Problems*

For more about the PPI Learning Hub, visit PPI's website at **ppi2pass.com**.

References

NCEES does not specify "codes and standards" in its lists of exam topics. For that reason, at least for the mechanical engineering PE exams, "codes and standards" seems to imply "knowledge about codes and standards," as opposed to "possession of and reference to the codes and standards" during the exam. The distinction is significant, because (without a specific list) it would be unreasonably expensive to purchase every code and standard affecting mechanical engineers. Among others, ASME, ASTM, ANSI, ASHRAE, SAE, NFPA, NEC, AGMA, EPA, OSHA, and other U.S. organizations publish numerous documents, as do Canada and the European Union (EU).

There are a few noteworthy exceptions: ASME Y14.5 (*Dimensioning and Tolerancing*); ASME *Boiler and Pressure Vessel Code* (BPVC) Sec. VIII, Div. 1; ASHRAE Standard 62.1 (*Ventilation for Acceptable Indoor Air Quality*); TEMA's *Standards of the Tubular Exchanger Manufacturers Association*; and OSHA CFR 29. Depending on your discipline, one or more of these publications could be valuable.

Inasmuch as fire protection is no longer a specific topic on the mechanical PE exams, none of the NFPA publications should be needed.

The following references were used to prepare these problems.

American Petroleum Institute. API 570-1998: *Piping Inspection Code: Inspection, Repair, Alteration, and Rerating of In-service Piping Systems.*

_____. API 682-2004/ISO 21049: *Pumps—Shaft Sealing Systems for Centrifugal and Rotary Pumps.*

American Society of Mechanical Engineers (ASME). B31.3-2006: *Process Piping.*

_____. *Boiler and Pressure Vessel Code.*

Keenan, Joseph H., et al., Steam Tables: *Thermodynamic Properties of Water Including Vapor, Liquid, and Solid Phases.* Krieger Publishing Company.

Shigley, Joseph E., et al., *Mechanical Engineering Design.* McGraw-Hill.

Volk, Michael W., *Pump Characteristics and Applications.* Marcel Dekker.

Nomenclature

a	acceleration	ft/sec^2	m/s^2		M	Mach number	–	–
A	area	ft^2	m^2		Ma	Mach number	–	–
BHP	boiler or brake horsepower	hp	hp		n	number of periods	–	–
					N	rotational speed	rpm	rpm
B	bulk modulus	lbft/ft^2	Pa		N	normal force	lbf	N
c	clearance	ft	m		p	pressure	lbf/ft^2	Pa
c	speed of sound	ft/sec	m/s		P	load	lbf	N
c_p	specific heat	Btu/lbm-°F	J/kg·°C		P	present worth	\$	\$
C	attachment factor	–	–		P	power	hp	W
C	coefficient	–	–		P	perimeter	ft	m
C	cost	\$	\$		q	heat	Btu	J
C_s	velocity of sound in fluid	ft/sec	m/s		Q	heat flow rate, heat transfer rate	Btu/hr	W
COP	coefficient of performance	–	–		Q	flow rate	gal/sec	L/s
CR	corrosion rate	mil/yr	n.a.		$\dot{Q}$	heat	Btu	J
d	diameter	ft	m		r	radius	ft	m
D	depreciation	\$	\$		R	resistance	Ω	Ω
D	diameter	ft	m		R	specific gas constant	ft-lbf/lbm-°R	J/kg·K
D_h	hydraulic diameter	ft	m		RCA	remaining corrosion allowance	ft	m
E	efficiency	–	–		Re	Reynolds number	–	–
E	quality factor	–	–		s	entropy	Btu/lbm-°F	J/kg·°C
f	friction factor	–	–		S	salvage value	\$	\$
F	correction factor	–	–		S	stress value	lbf/ft^2	Pa
F	force, load	lbf	N		S_n	expected salvage value in year n	\$	\$
F	future worth	\$	\$					
g	gravitational acceleration	ft/sec^2	m/s^2		SG	specific gravity	–	–
g_c	gravitational constant	ft-lbm/lbf-sec^2	n.a.		t	thickness	ft	m
					t	time	sec	s
h	enthalpy	Btu/lbm	J/kg		T	taxes	\$	\$
h	head, head loss, or height	ft	m		T	temperature	°F	°C
					U	overall heat transfer coefficient	Btu/hr-ft-°F	W/m^2·°C
i	effective rate of return	%	%		v	specific volume	ft^3/lbm	m^3/kg
I	current	A	A		v	velocity	ft/sec	m/s
k	ratio of specific heats	–	–		V	voltage	V	V
K	constant	–	–		V	volume	ft^3	m^3
K	minor loss coefficient	–	–		w	width	ft	m
L	length	ft	m		W	humidity ratio	–	–
m	mass	lbm	kg		W	mass flow rate	lbm/sec	kg/s
$\dot{m}$	mass flow rate	lbm/hr	kg/h		W	weight	lbf	N
M	Mach number	–	–		W	work	Btu/lbm	J/kg

W	power	hp	W
x	dimension	ft	m
x	quality	–	–
y	dimension	ft	m
y	mass fraction	–	–
z	elevation	ft	m

SYMBOLS

α	angle	deg	deg
γ	specific weight	lbf/ft^3	n.a.
ϵ	specific roughness	ft	m
η	efficiency	–	–
μ	absolute viscosity	$lbf\text{-}sec/ft^2$	Pa·s
μ	coefficient of friction	–	–
ν	kinematic viscosity	ft^2/sec	m^2/s
ρ	density	lbm/ft^3	kg/m^3
σ	stress	lbf/ft^2	Pa
τ	shear stress	lbf/ft^2	Pa
τ	torque	lbf/ft^2	Pa
x	dimension	ft^2	m
ω	angular velocity	rpm	rpm

Principles

PROBLEM 1

An F-22 Raptor is flying at an altitude of 50,000 ft at 1200 mph. The temperature at this altitude is 390°R. If air is treated as an ideal gas, the Raptor's Mach number is most nearly

- (A) 1.2
- (B) 1.6
- (C) 1.8
- (D) 2.2

Hint: Be sure to understand the definition of Mach number.

PROBLEM 2

The equation for airflow velocity over a 1 ft by 1 ft horizontal flat plate is empirically determined to be

$$v(x, y) = (5 - x)\left(\frac{y}{5} + \frac{y^2}{10} + \frac{y^3}{100}\right)$$

x and y are in feet, and the function $v(x, y)$ is in feet per second. x is measured from the midpoint of the leading edge of the plate, and y is measured up from the plate as shown. The viscosity, μ, is 122.8×10^{-7} lbm/sec-ft.

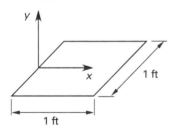

The total shear force acting on the upper surface of the plate is most nearly

- (A) 3.4×10^{-7} lbf
- (B) 3.8×10^{-7} lbf
- (C) 1.1×10^{-5} lbf
- (D) 1.2×10^{-5} lbf

Hint: Use Newton's law of viscosity.

PROBLEM 3

Water flows in a horizontal, square, cast-iron conduit measuring 2 in per side. The flow rate is 0.05 ft³/sec. The kinematic viscosity of the water is 1.08×10^{-5} ft²/sec, and the density of the water is 62.4 lbm/ft³. Assume all minor losses are incorporated in a minor loss coefficient, equal to 0.8. The expected pressure drop over 40 ft of conduit length is most nearly

- (A) 0.15 psi
- (B) 0.21 psi
- (C) 0.23 psi
- (D) 0.30 psi

Hint: The total energy loss is equal to the minor losses plus the friction loss.

PROBLEM 4

A pipeline carries sweet crude oil to a refinery at 14,000 gal/min. The diameter of the pipe is 24 in. The oil has a specific gravity of 0.86 and a kinematic viscosity of 4×10^{-5} ft²/sec. The flow inside the pipe is most nearly

- (A) laminar, with Re = 2000
- (B) laminar, with Re = 8500
- (C) turbulent, with Re = 41,000
- (D) turbulent, with Re = 500,000

Hint: Do not waste time worrying whether the 24 in diameter is nominal or actual.

PROBLEM 5

A plane is flying at Mach 1 at sea level. The temperature and pressure are at international standard atmospheric conditions. The maximum pressure on the nose can be calculated using compressible flow theory. If incompressible flow theory is used instead, the resulting error will most nearly be

(A) 10%

(B) 21%

(C) 27%

(D) 220%

Hint: The Bernoulli equation and isentropic tables are useful.

PROBLEM 6

An 800 gpm centrifugal pump is pumping a light hydrocarbon oil through a pipeline. The pump generates 700 ft of head, and the total efficiency of the pump and motor is 50%. The specific heat of the oil is 0.5 Btu/lbm-°F. The maximum possible temperature increase of the oil as it passes through the pump is most nearly

(A) 0.36°F

(B) 0.90°F

(C) 1.8°F

(D) 3.6°F

Hint: Be sure to include the efficiency in the calculation of the pump power.

PROBLEM 7

A boiler uses 220°F feedwater. It generates 85% quality saturated steam at a rate of 200,000 lbm/hr. The pressure of the steam is 1400 psia. The energy absorbed by the steam is most nearly

(A) 180×10^6 Btu/hr

(B) 200×10^6 Btu/hr

(C) 220×10^6 Btu/hr

(D) 280×10^6 Btu/hr

Hint: Think about enthalpies.

PROBLEM 8

In the piping system shown, 100 gpm of water is pumped from one tank to another.

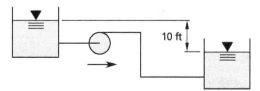

At this flow, the head loss due to piping and fittings is 90 ft. The pump has an efficiency of 60%, and it is driven by an electric motor with an efficiency of 90%. The fluid density is 50 lbm/ft^3. The power input to the motor is most nearly

(A) 0.38 hp

(B) 3.0 hp

(C) 3.4 hp

(D) 3.8 hp

Hint: Pump power depends on both pump head and flow rate.

PROBLEM 9

A pump moves water from a stagnant lake to a free jet 15 ft above the lake. Both locations are at atmospheric pressure. The pump can supply 20 ft of head with a water flow of 5 lbm/sec. The maximum velocity that this single pump can produce is most nearly

(A) 13 ft/sec

(B) 18 ft/sec

(C) 36 ft/sec

(D) 320 ft/sec

Hint: Use the extended Bernoulli equation.

PROBLEM 10

The condensate of a low-pressure steam system operating at 105 psig is discharged to an atmospheric vented vessel. The percentage of heat escaping in the flash steam is most nearly

(A) 14%

(B) 42%

(C) 73%

(D) 100%

Hint: The heat of vaporization of the condensate at atmospheric conditions will be needed.

PROBLEM 11

An intermediate-depth underwater mine designed for coastal defense consists of a smooth, hollow right circular cylinder that encases an MK42 torpedo. Because of the tether design, the system can be modeled as a double pendulum, as shown in the figure, with one pivot at the bottom of the case and one at the top of the anchor. The first two natural frequencies of oscillation for this particular system are 0.310 rad/sec and 1.81 rad/sec.

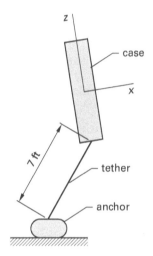

case properties	
length	122 in
outside diameter	23 in
inside diameter	22.25 in
weight	1526 lbf
CG location (measured from bottom of case)	71.2 in
I_x	728 ft-lbf-sec^2
I_y	728 ft-lbf-sec^2
I_z	unknown

When a cylinder is placed in a flowing liquid, it will experience a phenomenon known as vortex shedding. The vortex shedding frequency will be equal to the natural frequencies and is given in terms of the Strouhal number, S, which can be found from the equation shown.

$$S = \frac{nd}{v_0}$$

n is the vortex shedding frequency in hertz, d is the cylinder diameter, and v_0 is the free-stream velocity. The

Strouhal number can also be obtained from a graph of the Strouhal number versus the Reynolds number for the flow past a cylinder.

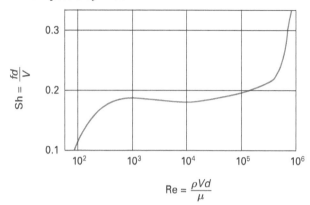

The current velocities that will cause the mine to oscillate at its two natural frequencies are most nearly

(A)   0.28 knots and 0.64 knots

(B)   0.48 knots and 2.8 knots

(C)   1.8 knots and 10 knots

(D)   3.4 knots and 20 knots

Hint: The Reynolds number will be needed in the solution, so it will be necessary to assume a reasonable number for the current velocity. For help in choosing such a number, look at the answer options.

PROBLEM 12

Water is drawn from a tank through a siphon as shown.

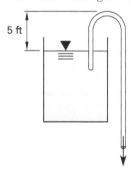

The water is at 70°F and atmospheric pressure (14.70 psia). The velocity of the water at the lower end of the siphon is 30 ft/sec. Friction losses are negligible. The pressure at the siphon's highest point is most nearly

(A)   0.44 psia

(B)   6.5 psia

(C)   8.7 psia

(D)   13 psia

Hint: Choose a zero-height reference that simplifies the Bernoulli equation.

PROBLEM 13

A schedule-80 pipe with a diameter of 10 in transports saturated water. The initial temperature of the water is 400°F, and the outside temperature is 66°F. The flow rate is 7230 lbm/hr. The inside and outside film coefficients can be assumed to be infinite. The pipe is insulated with calcium silicate so that temperature loss is limited to 1°F per 40 ft of pipe. The thermal conductivity of calcium silicate is 0.046 Btu-ft/hr-ft²-°F. The thermal resistance of the steel is negligible compared to that of the insulation. The minimum thickness of the insulation is most nearly

(A) 1.0 in

(B) 3.8 in

(C) 7.7 in

(D) 9.2 in

Hint: The thermal resistance of the steel is negligible compared to that of the insulation.

PROBLEM 14

An aluminum pot is used to boil water. The hollow pot handle is 7 in long, 0.5 in high, and 1 in wide with a uniform wall thickness of 1/16 in. The pot is in a 72°F room. The convection coefficient is 1.2 Btu/hr-ft²-°F.

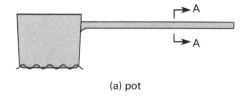

(a) pot

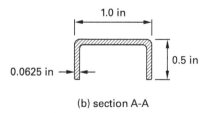

(b) section A-A

Model the pot handle as a fin, assuming the interior surfaces of the handle transfer negligible heat. The heat transfer rate from the handle is most nearly

(A) 14 Btu/hr

(B) 28 Btu/hr

(C) 48 Btu/hr

(D) 170 Btu/hr

Hint: A typical convection coefficient is 1.2 Btu/hr-ft²-°F.

PROBLEM 15

In the oil supply system shown, oil is pumped from the reservoir to a bearing via the manifold. The oil exiting the bearing is then collected and funneled back into the reservoir.

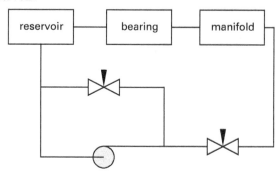

The bearing needs to be supplied with oil at 15 psi gauge. The primary functions of the two needle valves are to

   I.   control the pressure at the pump intake and the pressure at the manifold

  II.   control the flow rate to the manifold and the pressure at the manifold

 III.   control the flow rate to the pump intake and the flow rate to the manifold

 IV.   control the pressure at the pump intake and the flow rate to the manifold

(A) I only

(B) II only

(C) III only

(D) III and IV

Hint: Think about what happens if one needle valve is completely closed while the other is completely open.

PROBLEM 16

A jet of water from a stationary nozzle impinges on a movable block and vane, which are initially at rest on the ground. The total mass of the block and vane is 50 lbm, and the coefficient of static friction is 0.1. The vane has a turning angle of 30°. The water exits the nozzle and hits the vane at 150 ft/sec, with an area of contact of 0.005 ft². The friction between the vane and fluid is negligible.

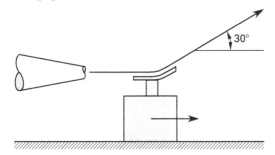

The initial acceleration of the mass is most nearly

(A)   0.27 ft/sec²

(B)   8.6 ft/sec²

(C)   19 ft/sec²

(D)   29 ft/sec²

Hint: Neglect the friction between the vane and the fluid.

PROBLEM 17

Air enters a diverging nozzle with a stagnation temperature of 240°F and a stagnation pressure of 200 psia. The speed of the air is Mach 1.8 just before it enters the nozzle. A shock stands at the entrance, and the exit area is three times larger than the throat area.

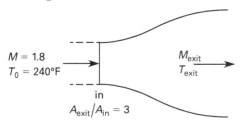

The exit Mach number is most nearly

(A)   0.17

(B)   0.26

(C)   0.6

(D)   2.8

Hint: An isentropic flow table and normal shock relationships will be needed.

PROBLEM 18

A 6 in schedule-40 pipe is used to transport 50,000 lbm/hr of 550°F steam at 500 psia. The Darcy friction factor for the flow is 0.02. The pressure drop per 100 ft of pipe is most nearly

(A)   2.2 lbf/in²

(B)   3.6 lbf/in²

(C)   6.1 lbf/in²

(D)   26 lbf/in²

Hint: A typical friction factor for turbulent flow in steel pipe is 0.02.

PROBLEM 19

A company is considering the purchase of a piece of equipment that costs $130,000. If purchased, the equipment will generate $45,000 of net profit each year for four years. This net profit represents all profits less operating and maintenance costs but before taxes. At the end of the four years, the equipment will be salvaged for $10,000, which will not be taxed. The company's tax rate is 45%. The minimum attractive rate of return is 8%. Using straight-line depreciation, the resulting present value is most nearly

(A)   −$5500

(B)   −$4000

(C) $4000

(D) $5500

Hint: A high present value indicates a more attractive option.

PROBLEM 20

The power output of an adiabatic steam turbine is 6700 hp, and the inlet and exit conditions are as shown.

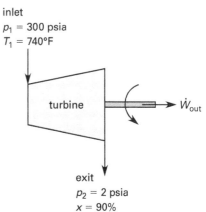

The mass flow rate of the steam is most nearly

(A) 4.6 lbm/sec

(B) 13 lbm/sec

(C) 17 lbm/sec

(D) 760 lbm/sec

Hint: Convert hp to Btu/hr.

PROBLEM 21

Which of the following can be done to increase the net positive suction head available to a pump?

   I. Use a larger diameter pipe.

   II. Partially close a throttling valve in the intake line.

   III. Pressurize the supply tank.

   IV. Increase the height of the supply tank above the pump.

(A) I and IV only

(B) II and IV only

(C) I, II, and IV

(D) I, III, and IV

Hint: This is a nonquantitative question. Start with the index of the *NCEES Handbook*.

PROBLEM 22

Which of the following are valid pipe fitting classifications?

   I. NPT

   II. SAE

   III. UNC

   IV. ASHRAE

(A) I and II

(B) I and III

(C) II and III

(D) III and IV

Hint: Do not confuse screw threads with pipe threads.

PROBLEM 23

The evaporator section of a heat recovery steam generator is made up of a drum boiler that generates the steam, a tube bank heat exchanger, and pipes that carry the water and steam. Preheated water arrives in the drum at 580°F. The drum operates at 1400 psia. It produces 95% quality steam at a rate of 540,000 lbm/hr. Each tube in the tube bank has a useful heat-exchange surface of 5.3 ft^2 and an overall coefficient of heat transfer of 218.2 Btu/hr-ft^2-°F. The average temperature difference between the water in the tube bank and the flue gas is 32°F. Because of heat losses, only 92% of the heat transmitted to the drum contributes to the generation of steam. The number of tubes needed to maintain the required steam production is most nearly

(A) 7500 tubes

(B) 8800 tubes

(C) 9200 tubes

(D) 280,000 tubes

Hint: Find the heat transfer rate needed to produce the steam.

PROBLEM 24

An axial flow pump running at sea level is used to transfer water from an open-air tank to a higher level. The temperature and pressure are at standard atmospheric conditions. The friction loss in the suction pipe is 5 ft. The net positive suction head available is 10 ft. The suction lift is most nearly

(A) 10 ft

(B) 18 ft

(C) 26 ft

(D) 30 ft

Hint: Be sure to use units of "feet absolute."

PROBLEM 25

A boiler horsepower is the amount of power needed to convert 34.5 lbm/hr of feedwater at 212°F and atmospheric pressure to dry, saturated steam at the same temperature and pressure. The equivalent amount of Btu/hr is most nearly

(A) 34,000 Btu/hr

(B) 40,000 Btu/hr

(C) 42,000 Btu/hr

(D) 98,000 Btu/hr

Hint: h_{fg} will be useful here.

PROBLEM 26

Section I of the ASME *Boiler and Pressure Vessel Code* states that the relieving capacities of steam relief safety valves are given by the equation shown.

$$W_{th} = 51.5Ap$$

The theoretical flow, W_{th}, is in lbm/hr. The nozzle throat area, A, is in square inches. The pressure, p, for 3% accumulation is calculated using the equation shown.

$$p = 1.03(\text{set pressure}) + 14.7 \ \frac{\text{lbf}}{\text{in}^2}$$

ASME code, however, says to use either the preceding equation or the following, whichever gives a greater value for p.

$$p = \text{set pressure} + 2 \ \frac{\text{lbf}}{\text{in}^2} + 14.7 \ \frac{\text{lbf}}{\text{in}^2}$$

The smallest set pressure for which the first equation will give a larger pressure is most nearly

(A) 52 psi

(B) 67 psi

(C) 82 psi

(D) 96 psi

Hint: This problem can be solved by using algebra or by graphing.

PROBLEM 27

A nozzle of SA-106 grade B material abuts a seamless section of a pressure vessel. The inner diameter of the nozzle is 4.5 in. The pressure vessel has a maximum allowable working pressure of 250 psi and a design temperature of 500°F. The pressure vessel's inside diameter is 68 in, and its walls are 0.625 in thick. The required

nozzle reinforcement area is calculated using the equation shown.

$$A_r = Dt_{\text{req}}$$

D is the finished diameter of the circular opening in the shell, and t_{req} is the required thickness of the shell based on the circumferential stress. From the ASME *Boiler and Pressure Vessel Code*, Sec. VIII, Div. I, the required thickness of the shell is found using the equation shown.

$$t_{\text{req}} = \frac{pr}{SE - 0.6p}$$

p is the pressure, and S is the allowable stress (from ASME Sec. II, $S = 17.1$ ksi for SA-106 grade B). E is the joint efficiency ($E = 1.0$ for a seamless shell). If the nozzle provides 0.518 in^2 of reinforcement, the amount of additional reinforcement needed is most nearly

(A) 1.0 in^2

(B) 1.2 in^2

(C) 2.7 in^2

(D) 3.4 in^2

Hint: ASME *Boiler and Pressure Vessel Code*, Sec. VIII, Div. I says the reinforcing area of the shell, A_s, plus the reinforcing area of the nozzle, A_n, must at least equal the required reinforcement area, A_r. $A_s + A_n \geq A_r$.

PROBLEM 28

The remaining corrosion allowance (RCA) of a section of 3 in seamless schedule-40 SA-106 grade B pipe must be determined. The pipe is in a crude distillation unit that operates at 250 psi and 600°F. The pipe was installed in 1990, and subsequent ultrasonic thickness readings are shown.

year	thickness (in)
1995	0.208
1998	0.200
current year	0.175

ASME B31.3 says that the RCA is equal to the actual thickness less the minimum required thickness, and that the equation shown can be used to determine the minimum required thickness.

$$t_{\text{min}} = \frac{pD_o}{2SE}$$

p is the pressure, D_o is the outside pipe diameter, S is the allowable stress (from ASME Sec. II, $S = 17.1$ ksi for SA-106 grade B), and E is the quality factor ($E = 1$ for a seamless pipe).

The remaining corrosion allowance is most nearly

(A) 0.12 in

(B) 0.15 in

(C) 0.16 in

(D) 0.19 in

Hint: Begin by finding the outer diameter.

PROBLEM 29

The coefficient of friction in lightly loaded, full journal bearing is given by the equation shown.

$$f = \pi^2 \left(\frac{\mu N}{P} \right) \left(\frac{r}{c} \right)$$

What units must the dynamic viscosity be given in to find the coefficient of friction?

(A) Pa·sec

(B) lbm-sec/ft

(C) poise

(D) reyn

Hint: This coefficient of friction comes from Petroff's law.

PROBLEM 30

A refrigeration cycle uses HFC-134a as the refrigerant. The condensing temperature in the cycle is 120°F, and the evaporation temperature is 0°F. The compression ratio is most nearly

(A) 0.11

(B) 5.6

(C) 8.8

(D) 15

Hint: The R134a saturated liquid and saturated vapor table will be useful.

PROBLEM 31

The feedwater temperature in a boiler is 280°F. The quality of 1200 psia steam required to absorb 800 Btu/lbm from the boiler is most nearly

(A) 37%

(B) 53%

(C) 78%

(D) 85%

Hint: Do not forget about the energy in the feedwater.

PROBLEM 32

Air is flowing at a rate of 1 lbm/sec. The temperature is 70°F, and the pressure is 14.7 psia. The air is to be compressed to 200 psia. Which of the following statements are true?

I. An isentropic process would require the least amount of power.

II. An adiabatic process must also be an isentropic process.

III. The amount of power required for an isentropic process would be within 10% of 200 hp.

(A) I only

(B) II only

(C) III only

(D) I and III only

Hint: An ideal isothermal process uses cooling between staged compressors.

PROBLEM 33

A flat, 3 ft diameter, unstayed head is to be designed for a life of 30 years and a design pressure of 150 psi. ASME *Boiler and Pressure Vessel Code*, Sec. VIII, Div. 1 says the minimum thickness for an unstayed head is found from the equation shown.

$$t_{\min} = D \sqrt{\frac{Cp}{SE}}$$

For this design, the maximum allowable stress, S, is

17,500 psi. The attachment factor, C, is 0.20, and the joint efficiency, E, is 1. The corrosion rate is 8 mil/yr. The design thickness of the head is most nearly

(A) 0.24 in

(B) 0.36 in

(C) 1.7 in

(D) 3.6 in

Hint: Remember this head can't be less than t_{min} for its entire life.

PROBLEM 34

A house contains air at 68°F and 75% relative humidity. The minimum window temperature needed to keep moisture from condensing on the inner surfaces of the windows is most nearly

(A) 25°F

(B) 40°F

(C) 50°F

(D) 60°F

Hint: A psychrometric chart will be useful.

PROBLEM 35

A vehicle drives at 50 mph into a 10 mph headwind with a cartop cargo carrier on the roof. If the effective frontal area of the cargo carrier is 480 in^2, the additional power needed to overcome the drag force due to the cargo carrier is most nearly

(A) 1.9 hp

(B) 3.5 hp

(C) 6.0 hp

(D) 12 hp

Hint: Use a drag coefficient of 1.25.

PROBLEM 36

An air-standard diesel engine has a compression ratio of 16:1. The air pressure at the end of the compression process is most nearly

(A) 110 psi

(B) 240 psi

(C) 540 psi

(D) 710 psi

Hint: In an air-standard engine, compression and expansion processes are assumed to be isentropic.

PROBLEM 37

Air is held in a reservoir at 20°C and 500 kPa (absolute). The Mach number at which the air exits to the atmosphere through a hole is most nearly

(A) 0.01

(B) 1.7

(C) 2.1

(D) 4.4

Hint: Treat the hole as a converging-diverging duct.

PROBLEM 38

A carbon dioxide sequestration pipeline has an inside diameter of 6 in. At one point in the pipeline, the CO_2 is at 80°F and 40 psia. The velocity is 20 ft/sec. Further down the pipeline, the CO_2 passes a point where the pressure is 30 psia at 100°F. The flow at the downstream point is most nearly

(A) 2.9 ft^3/sec

(B) 5.4 ft^3/sec

(C) 6.5 ft^3/sec

(D) 28 ft^3/sec

Hint: The mass flow rate is constant, but the volumetric flow rate is not.

PROBLEM 39

A new check valve is being developed for a city's water main pipeline system. The prototype valve has 2 ft inlet and exit diameters that carry water at a rate of 40 ft³/sec. A model is being created to test this valve using a flow rate of 5 ft³/sec. The inlet and exit diameters of the model need to be most nearly

(A) 0.25 in

(B) 3.0 in

(C) 8.5 in

(D) 16 in

Hint: Use dynamic similarity.

PROBLEM 40

The pipe assembly shown is made from schedule-40 pipe. At location C, the nominal diameter of the pipe is 0.75 in, and the static pressure is 40 psi. At locations A and B, the nominal diameter of the pipe is 0.5 in, and the water flows out of the pipe assembly into atmospheric air. At location A, the velocity of the water is 12 ft/sec, and at location B, the velocity of the water is 25 ft/sec. The water is at 60°F. Neglect the weight of the pipe and the water in the pipe.

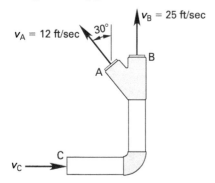

The horizontal force exerted on the elbow at location C needed to hold the pipe assembly in equilibrium is most nearly

(A) 3.5 lbf

(B) 16 lbf

(C) 21 lbf

(D) 25 lbf

Hint: A control volume will be useful here.

PROBLEM 41

The water velocity in a 1 in, schedule-40 pipe with a rapid-closing valve is 3 ft/sec. The water is at standard temperature, and the initial pressure in the pipe is 60

psi. The bulk modulus of water is 300,000 psi. When the valve is closed, the maximum pressure that will result is most nearly

(A) 60 psi

(B) 130 psi

(C) 200 psi

(D) 250 psi

Hint: The speed of sound in water will be needed.

PROBLEM 42

A pump curve is shown. The flow rate at the pump's operating point is 400 gal/min.

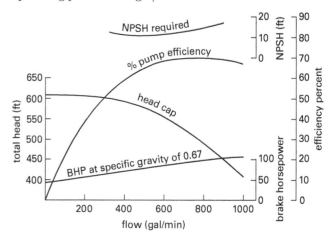

The amount of pump head needed at the pump shaft is most nearly

(A) 840 ft

(B) 980 ft

(C) 1100 ft

(D) 1500 ft

Hint: The operating point is important.

PROBLEM 43

An ideal Rankine cycle and an incomplete data table are shown.

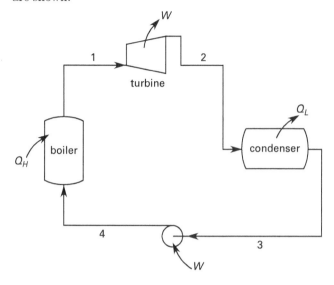

state	pressure (psi)	temperature (°F)	enthalpy (Btu/lbm)	entropy (Btu/ lbm-°R)
1	500	700	1357	1.612
2	2			
3			94	
4			96	

The heat removed by the condenser per pound mass is most nearly

(A) 94 Btu/lbm

(B) 840 Btu/lbm

(C) 1020 Btu/lbm

(D) 1260 Btu/lbm

Hint: The steam quality at state 2 will be needed.

PROBLEM 44

For a combined gas turbine/vapor power cycle, which two statements are true?

  I. The thermal efficiency of the combined cycle is greater than the thermal efficiency of either cycle working individually.

  II. Gas turbine power cycles typically operate at lower temperatures than steam power cycles.

  III. A binary vapor power cycle is classified as a combined power cycle.

  IV. The practical upper limit for the efficiency of a combined power cycle is considered to be approximately 81%.

(A) I and III

(B) II and IV

(C) II and III

(D) III and IV

Hint: One of the true statements is not intuitive.

SOLUTION 1

The Mach number of an object is the ratio of the object's speed to the speed of sound in the medium it is moving through. The equation for the speed of sound, c, in an ideal gas is

Mach Number

$$c = \sqrt{kRT}$$

Since air is treated as an ideal gas, the ratio of specific heats, k, and the specific gas constant, R, are both constant. For air, the ratio of specific heats is $k = 1.40$, and the specific gas constant for air is 53.35 ft-lbf/lbm-°R. [Thermal and Physical Properties of Ideal Gases (at Room Temperature)]

For the given temperature of 390°R, the speed of sound in air at 50,000 ft is

Mach Number

$$c = \sqrt{kRT} = \sqrt{kg_c RT}$$

$$= \sqrt{\begin{array}{c}(1.4)\left(32.17\ \dfrac{\text{lbm-ft}}{\text{lbf-sec}^2}\right)\\ \times\left(53.35\ \dfrac{\text{ft-lbf}}{\text{lbm-°R}}\right)(390°\text{R})\end{array}}$$

$$= 968\ \text{ft/sec}$$

The Mach number is

Mach Number

$$\text{Ma} = \frac{\text{v}}{c}$$

$$= \frac{\left(\dfrac{\left(1200\ \dfrac{\text{mi}}{\text{hr}}\right)\left(5280\ \dfrac{\text{ft}}{\text{mi}}\right)}{3600\ \dfrac{\text{sec}}{\text{hr}}}\right)}{968\ \dfrac{\text{ft}}{\text{sec}}}$$

$$= 1.82\ \ (1.8)$$

The answer is (C).

Why Other Options Are Wrong

(A) This incorrect answer results when the speed of sound at sea level (1130 ft/sec) is used instead of the Raptor's velocity to compute the Mach number.

(B) This incorrect answer results when the speed of sound at sea level (1130 ft/sec) is used instead of the speed of sound at 50,000 ft to compute the Mach number.

(D) This incorrect answer results when the ratio of specific heats is neglected in calculating the speed of sound.

SOLUTION 2

Newton's law of viscosity applied at the surface (where y equals zero) is

Stress, Pressure, and Viscosity

$$\tau = \mu\,\frac{d\text{v}}{dy}$$

$$= \frac{\mu}{g_c}\left.\frac{\partial\text{v}(x,y)}{\partial y}\right|_{y=0}$$

$$= \left(\frac{\mu}{g_c}\right)(5-x)\left(\frac{1}{5}+\frac{y}{5}+\frac{3y^2}{100}\right)\Bigg|_{y=0}$$

$$= \left(\frac{\mu}{g_c}\right)(5-x)\left(\frac{1}{5}\ \frac{1}{\text{sec}}\right)$$

$\text{v}(x,\ y)$ has units of ft/sec, so the derivative of $\text{v}(x,\ y)$ with respect to the normal distance, y, has units of sec^{-1}, and the shear stress, τ, has units of lbf/ft². After integrating, the unit area in square feet can multiplied into the result, giving an answer in pounds-force.

The shear force is obtained by integrating the shear stress over the area of the plate. The differential area, dA, is equal to $dx\,dz$; x goes from 0 ft to 1 ft, and z goes from −0.5 ft to 0.5 ft.

$$F = \int \tau_{y=0}\,dA$$

$$= \int_{z=-0.5\,\text{ft}}^{0.5\,\text{ft}}\int_{x=0\,\text{ft}}^{1\,\text{ft}}\left(\frac{\mu}{(5\ \text{ft-sec})g_c}\right)(5\ \text{ft}-x)\,dx\,dz$$

$$= \int_{z=-0.5\,\text{ft}}^{0.5\,\text{ft}}dz\int_{x=0\,\text{ft}}^{1\,\text{ft}}\left(\frac{\mu}{(5\ \text{ft-sec})g_c}\right)(5\ \text{ft}-x)\,dx$$

$$= (z\vert_{-0.5\,\text{ft}}^{0.5\,\text{ft}})\left(\left(\frac{\mu}{(5\ \text{ft-sec})g_c}\right)\left((5\ \text{ft})x-\frac{x^2}{2}\right)\right)\Bigg|_{0\,\text{ft}}^{1\,\text{ft}}$$

$$= (1\ \text{ft})\left(\frac{\mu}{(5\ \text{ft-sec})g_c}\right)(4.5\ \text{ft}^2)$$

$$= \left(\frac{\mu}{g_c}\right)\left(0.9\ \frac{\text{ft}^2}{\text{sec}}\right)$$

$$= \left(\frac{122.8\times10^{-7}\ \dfrac{\text{lbm}}{\text{sec-ft}}}{32.17\ \dfrac{\text{lbm-ft}}{\text{lbf-sec}^2}}\right)\left(0.9\ \frac{\text{ft}^2}{\text{sec}}\right)$$

$$= 3.44\times10^{-7}\ \text{lbf}\ \ (3.4\times10^{-7}\ \text{lbf})$$

The answer is (A).

Why Other Options Are Wrong

(B) This incorrect answer results when the shear stress is evaluated for $x = 0$ and $y = 0$, and the result is multiplied by the area of the plate (1 ft^2).

(C) This incorrect answer results when the viscosity is not divided by g_c.

(D) This incorrect answer results when the shear stress is evaluated for $x = 0$ and $y = 0$, and the viscosity is not divided by g_c.

SOLUTION 3

The Bernoulli equation is

Bernoulli Equation

$$\frac{p_1}{\rho g} + z_1 + \frac{v_1^2}{2g} = \frac{p_2}{\rho g} + z_2 + \frac{v_2^2}{2g} + h_f$$

When density, ρ, is in lbm/ft^3, the Bernoulli equation can be expressed as

$$\frac{p}{\rho} + \frac{g_z}{g_c} + \frac{v}{2g_c} = \frac{p}{\rho} + \frac{g_z}{g_c} + \frac{v}{2g_c} + \frac{g_h}{g_c}$$

For a horizontal pipe, z_1 and z_2 are equal. The pipe has a constant cross-section and is continuous, so v_1 and v_2 are equal. Simplifying and solving for the pressure drop gives

$$p_1 - p_2 = \frac{\rho g h_{losses}}{g_c}$$

The total losses are equal to the minor losses plus the friction loss.

$$p_1 - p_2 = \frac{\rho g (h_m + h_f)}{g_c}$$

To calculate the minor losses, the average velocity in the pipe is needed. For a square duct with side length L,

Continuity Equation

$$Q = Av$$

$$v = \frac{Q}{A} = \frac{Q}{L^2} = \frac{0.05 \ \frac{ft^3}{sec}}{\left(\frac{2 \ in}{12 \ \frac{in}{ft}}\right)^2} = 1.80 \ ft/sec$$

The minor losses can be calculated from

Valve and Fittings Losses

$$\Delta h_m = K\left(\frac{v^2}{2g}\right)$$

$$= (0.8)\left(\frac{\left(1.80 \ \frac{ft}{sec}\right)^2}{(2)\left(32.17 \ \frac{ft}{sec^2}\right)}\right)$$

$$= 0.04025 \ ft$$

To calculate the frictional losses, the friction factor and the pipe diameter are needed. For a noncircular pipe, the hydraulic diameter, D_h, must be used.

Hydraulic Diameter

$$D_h = \frac{4A}{P}$$

$$= \frac{4L^2}{4L}$$

$$= L$$

$$= \frac{2 \ in}{12 \ \frac{in}{ft}}$$

$$= 0.1667 \ ft$$

To find the friction factor, the Reynolds number and relative roughness for the flow are needed. The Reynolds number is

Reynolds Number

$$Re = \frac{v D_h}{\nu}$$

$$= \frac{\left(1.80 \ \frac{ft}{sec}\right)(0.1667 \ ft)}{1.08 \times 10^{-5} \ \frac{ft^2}{sec}}$$

$$= 2.78 \times 10^4$$

A cast-iron pipe has a specific roughness in the range 0.0006 ft to 0.003 ft; assume an intermediate value of 0.0018 ft. [Moody Diagram (Stanton Diagram)]

The relative roughness is

$$\frac{\epsilon}{D_h} = \frac{0.0018 \ ft}{0.1667 \ ft} = 0.01080$$

For these values, a Moody friction factor chart gives a friction factor of approximately 0.04. [Moody Diagram (Stanton Diagram)]

Use the Darcy-Weisbach equation to calculate the frictional losses. Let point 1 be 40 ft upstream from point 2.

Head Loss Due to Flow: Darcy-Weisbach Equation

$$h_f = f\left(\frac{L}{D}\right)\left(\frac{v^2}{2g}\right)$$

$$= (0.04)\left(\frac{40 \text{ ft}}{0.1667 \text{ ft}}\right)\left(\frac{\left(1.80 \frac{\text{ft}}{\text{sec}}\right)^2}{(2)\left(32.17 \frac{\text{ft}}{\text{sec}^2}\right)}\right)$$

$$= 0.4829 \text{ ft}$$

The pressure drop is

$$p_1 - p_2 = \frac{\rho g(h_m + h_f)}{g_c}$$

$$= \frac{\left(62.4 \frac{\text{lbm}}{\text{ft}^3}\right)\left(32.17 \frac{\text{ft}}{\text{sec}^2}\right)}{\left(32.17 \frac{\text{lbm-ft}}{\text{lbf-sec}^2}\right)\left(12 \frac{\text{in}}{\text{ft}}\right)^2}$$

$$= 0.2267 \text{ lbf/in}^2 \quad (0.23 \text{ psi})$$

The answer is (C).

Why Other Options Are Wrong

(A) This incorrect answer results when the relative roughness is calculated without converting the pipe's hydraulic diameter to feet.

(B) This incorrect answer results when the minor losses are neglected.

(D) This incorrect answer results when the area of the pipe is calculated as a circular cross-section of diameter 2 in.

SOLUTION 4

The magnitude of the Reynolds number indicates whether the flow is laminar or turbulent. To find the Reynolds number, the average velocity of the flow inside the pipe is needed. Use the continuity equation, and solve for the average velocity.

Continuity Equation

$$Q = vA$$

$$v = \frac{Q}{A} = \frac{Q}{\left(\frac{\pi D^2}{4}\right)} = \frac{4Q}{\pi D^2}$$

$$= \frac{(4)\left(\frac{14,000 \frac{\text{gal}}{\text{min}}}{\left(7.481 \frac{\text{gal}}{\text{ft}^3}\right)\left(60 \frac{\text{sec}}{\text{min}}\right)}\right)}{\pi\left(\frac{24 \text{ in}}{12 \frac{\text{in}}{\text{ft}}}\right)^2}$$

$$= 9.9 \text{ ft/sec}$$

The Reynolds number is

Reynolds Number

$$Re = \frac{vD}{\nu}$$

$$= \frac{\left(9.9 \frac{\text{ft}}{\text{sec}}\right)\left(\frac{24 \text{ in}}{12 \frac{\text{in}}{\text{ft}}}\right)}{4 \times 10^{-5} \frac{\text{ft}^2}{\text{sec}}}$$

$$= 500,000$$

The Reynolds number is greater than 12,000, so the flow is turbulent. [Reynolds Number]

The answer is (D).

Why Other Options Are Wrong

(A) This incorrect answer results when a flow rate of 1400 is mistakenly used, πD is used for the area, and the inches-to-feet conversion is neglected when the flow velocity is calculated.

(B) This incorrect answer results when the number of gallons per minute is divided by 3600 instead of 60 when calculating v.

(C) This incorrect answer results when the inches-to-feet conversion is neglected.

SOLUTION 5

For incompressible flow, Bernoulli's equation is needed.

Bernoulli Equation

$$\frac{p_1}{\rho g} + z_1 + \frac{v_1^2}{2g} = \frac{p_2}{\rho g} + z_2 + \frac{v_2^2}{2g} + h_f$$

$$\frac{p_1}{\rho} + \frac{v_1^2}{2g_c} + \frac{gz_1}{g_c} = \frac{p_2}{\rho} + \frac{v_2^2}{2g_c} + \frac{gz_2}{g_c}$$

For a streamline connecting a point away from the plane (point 1) to the stagnation point at the nose (point 2), z_1 is equal to z_2, and v_2 is equal to zero. Reducing and solving for p_2 gives

$$\frac{p_1}{\rho} + \frac{v_1^2}{2g_c} = \frac{p_2}{\rho}$$

$$p_2 = p_1 + \frac{\rho v_1^2}{2g_c}$$

Density can be calculated from the equation of state for ideal gases. p_1 is 14.70 psia at standard atmospheric conditions. [Standard Dry Air Conditions at Sea Level]

For air, the gas constant, R, is 53.35 ft-lbf/lbm-°R. [Thermal and Physical Properties of Ideal Gases (at Room Temperature)]

For air at sea level, the temperature is 69°F, which is equal to 529°R. [Standard Dry Air Conditions at Sea Level]

Equations of State (EOS)

$$p = \frac{RT}{v}$$

$$\rho = \frac{1}{v} = \frac{p}{RT}$$

$$= \frac{\left(\left(14.70 \ \frac{\text{lbf}}{\text{in}^2}\right)\left(12 \ \frac{\text{in}}{\text{ft}}\right)^2\right)}{\left(53.35 \ \frac{\text{ft-lbf}}{\text{lbm-°R}}\right)(529°\text{R})}$$

$$= 0.075 \ \text{lbm/ft}^3$$

Because the plane is flying at Mach 1, the local speed of sound is needed to find v_1. For air, the ratio of specific heats is $k = 1.40$. [Thermal and Physical Properties of Ideal Gases (at Room Temperature)]

$$c = \sqrt{kRT} = \sqrt{kg_c RT}$$

$$= \sqrt{\frac{(1.40)\left(32.17 \ \frac{\text{lbm-ft}}{\text{lbf-sec}^2}\right)}{\times \left(53.35 \ \frac{\text{ft-lbf}}{\text{lbm-°R}}\right)(529°\text{R})}}$$

$$= 1127.9 \ \text{ft/sec}$$

Substituting the values into the simplified Bernoulli equation gives

$$p_{2,i} = p_1 + \frac{\rho v_1^2}{2g_c}$$

$$= 14.70 \ \frac{\text{lbf}}{\text{in}^2}$$

$$+ \frac{\left(0.075 \ \frac{\text{lbm}}{\text{ft}^3}\right)\left(1127.9 \ \frac{\text{ft}}{\text{sec}}\right)^2}{(2)\left(32.17 \ \frac{\text{lbm-ft}}{\text{lbf-sec}^2}\right)\left(12 \ \frac{\text{in}}{\text{ft}}\right)^2}$$

$$= 24.98 \ \text{lbf/in}^2$$

From a table of isentropic compressible flow functions, for compressible flow at Mach 1, the ratio of static to stagnation or total pressure, p/p_0, is 0.5283. [One-Dimensional Isentropic Compressible-Flow Functions $k = 1.4$]

For compressible flow, the maximum stagnation pressure is

$$\frac{p}{p_0} = \frac{p}{p_{2,c}}$$

$$= 0.5283$$

$$p_{2,c} = \frac{p}{0.5283} = \frac{14.70 \ \frac{\text{lbf}}{\text{in}^2}}{0.5283} = 27.82 \ \text{lbf/in}^2$$

The fractional error is

$$\varepsilon = \frac{|p_{2,c} - p_{2,i}|}{p_{2,c}}$$

$$= \frac{\left|27.82 \ \frac{\text{lbf}}{\text{in}^2} - 24.98 \ \frac{\text{lbf}}{\text{in}^2}\right|}{27.82 \ \frac{\text{lbf}}{\text{in}^2}}$$

$$= 0.102 \quad (10\%)$$

The answer is (A).

Why Other Options Are Wrong

(B) This incorrect answer results when k is omitted in calculating v_1.

(C) This incorrect answer results when ρv_1^2 is divided g_c instead of $2g_c$.

(D) This incorrect answer results when p is multiplied by 0.5283 instead of being divided by it.

SOLUTION 6

The pump power equation is

Pump Power Equation

$$\dot{W} = \frac{Q\rho g h}{\eta_t}$$

By the continuity equation, $\dot{m} = Q\rho$. $\dot{W}$ is in ft-lbf/sec, so the gravitational constant must be used to convert pounds-mass to pounds-force. The pump power equation can then be written as

$$\dot{W} = \frac{\dot{m}gh}{\eta_t g_c}$$

The equation for the heat flow rate is

Compressors

$$\dot{W}_{\text{comp}} = \dot{m}c_p(T_e - T_i)$$

The maximum possible temperature increase will occur when all the pump power goes into heating the fluid. Equating the heat flow rate equation to the pump power equation gives

$$\frac{\dot{m}gh}{\eta_t g_c} = \dot{m}c_p\Delta T$$

Solve for the temperature rise, converting the heat flow rate from Btu to ft-lbf.

$$\Delta T = \frac{hg}{c_p\eta_t g_c}$$

$$= \frac{(700\ \text{ft})\left(32.17\ \dfrac{\text{ft}}{\text{sec}^2}\right)}{\left(\left(0.5\ \dfrac{\text{Btu}}{\text{lbm-}^\circ\text{F}}\right)\left(778\ \dfrac{\text{ft-lbf}}{\text{Btu}}\right)\right)}$$

$$\times (0.5)\left(32.17\ \frac{\text{lbm-ft}}{\text{lbf-sec}^2}\right)$$

$$= 3.6^\circ\text{F}$$

The answer is (D).

Why Other Options Are Wrong

(A) This incorrect answer results when a head of 70 ft is used instead of 700 ft.

(B) This incorrect answer results when, in the calculation of the brake horsepower, the water horsepower is multiplied by the efficiency of the pump instead of dividing by it.

(C) This incorrect answer results when the efficiency of the pump is neglected.

SOLUTION 7

The enthalpy of the 220°F feedwater is $h_{\text{feed}} = 188.27$ Btu/lbm. [Properties of Saturated Water and Steam (Temperature) - I-P Units]

The enthalpy of the 85% quality, 1400 psia saturated steam can be calculated from the values of h_f and h_{fg}. From steam tables, [Properties of Saturated Water and Steam (Pressure) - I-P Units]

$$h_f = 598.85\ \text{Btu/lbm}$$

$$h_{\text{fg}} = 575.60\ \text{Btu/lbm}$$

The enthalpy of the 85% quality steam is

Properties for Two-Phase (Vapor-Liquid) Systems

$$h = h_f + xh_{\text{fg}}$$

$$= 598.85\ \frac{\text{Btu}}{\text{lbm}} + (0.85)\left(575.60\ \frac{\text{Btu}}{\text{lbm}}\right)$$

$$= 1088\ \text{Btu/lbm}$$

The energy absorbed by the steam is equal to the heat flow rate.

$$q = \dot{m}\Delta h$$

$$= \left(200{,}000\ \frac{\text{lbm}}{\text{hr}}\right)\left(1088\ \frac{\text{Btu}}{\text{lbm}} - 188.27\ \frac{\text{Btu}}{\text{lbm}}\right)$$

$$= 180 \times 10^6\ \text{Btu/hr}$$

The answer is (A).

Why Other Options Are Wrong

(B) This incorrect answer results when the enthalpy of 120°F feedwater is used instead of the enthalpy of 220°F feedwater.

(C) This incorrect answer results when the initial feedwater enthalpy is not subtracted from the final steam enthalpy.

(D) This incorrect answer results when h_g is used instead of h_{fg}.

SOLUTION 8

The head the pump must provide is the friction loss minus the elevation change, because the liquid level in the discharge tank is lower than that of the supply tank.

$$h_{\text{pump}} = h_f - \Delta z = 90 \text{ ft} - 10 \text{ ft}$$
$$= 80 \text{ ft}$$

Convert the flow to cubic feet per minute. [Measurement Relationships]

$$Q = \frac{100 \ \dfrac{\text{gal}}{\text{min}}}{7.481 \ \dfrac{\text{gal}}{\text{ft}^3}} = 13.37 \text{ ft}^3/\text{min}$$

The power input to the motor is

Pump Power Equation

$$\dot{W} = \frac{Q\rho gh}{\eta_t}$$

$$= \frac{Q\rho gh_{\text{pump}}}{\eta_{\text{pump}}\eta_{\text{motor}}}$$

$$= \frac{\left(13.37 \ \dfrac{\text{ft}^3}{\text{min}}\right)\left(50 \ \dfrac{\text{lbm}}{\text{ft}^3}\right)}{(0.60)(0.90)\left(32.17 \ \dfrac{\text{lbm-ft}}{\text{lbf-sec}^2}\right)}$$
$$\times \left(32.17 \ \dfrac{\text{ft}}{\text{sec}^2}\right)(80 \ \text{ft})$$
$$\times \left(33{,}000 \ \dfrac{\text{ft-lbf}}{\text{hp-min}}\right)$$

$$= 3.001 \text{ hp} \quad (3.0 \text{ hp})$$

The answer is (B).

Why Other Options Are Wrong

(A) This incorrect answer results when friction loss is neglected.

(C) This incorrect answer results when the elevation change is neglected.

(D) This incorrect answer results when the elevation change is added to the friction loss instead of being subtracted.

SOLUTION 9

Use the Bernoulli equation.

Bernoulli Equation

$$\frac{p_1}{\rho g} + z_1 + \frac{\text{v}_1^2}{2g} = \frac{p_2}{\rho g} + z_2 + \frac{\text{v}_2^2}{2g} + h_f$$

Including a term for the head added by the pump gives

$$\frac{p_1}{\rho g} + z_1 + \frac{\text{v}_1^2}{2g} + h_{\text{pump}} = \frac{p_2}{\rho g} + z_2 + \frac{\text{v}_2^2}{2g} + h_f$$

Define location 1 as the stagnant lake surface and location 2 as the free jet's exit. The velocity at the lake surface, v_1, is negligible. Pressure is atmospheric at both the lake surface and the free jet's exit, so the pressure terms cancel. To obtain the maximum velocity, assume that friction loss, h_f, is zero. Simplifying the equation and solving for velocity at the free jet exit gives

$$\text{v}_2 = \sqrt{(2g)\big((z_1 - z_2) + h_{\text{pump}}\big)}$$

$$= \sqrt{(2)\left(32.17 \ \dfrac{\text{ft}}{\text{sec}^2}\right)(-15 \text{ ft} + 20 \text{ ft})}$$

$$= 17.94 \text{ ft/sec} \quad (18 \text{ ft/sec})$$

The answer is (B).

Why Other Options Are Wrong

(A) This incorrect answer results when the velocities in the Bernoulli equation are divided by g_c instead of $2g_c$.

(C) This incorrect answer results when the 15 ft elevation change is neglected.

(D) This incorrect answer results when the square root in the equation for v_2 is omitted.

SOLUTION 10

To use the steam tables, the pressures must be absolute.

$$p_1 = 105 \ \frac{\text{lbf}}{\text{in}^2} + 14.70 \ \frac{\text{lbf}}{\text{in}^2} = 119.7 \text{ lbf/in}^2 \approx 120 \text{ lbf/in}^2$$

For an absolute pressure of approximately 120 lbf/in², the enthalpy h_f is 312.55. [Properties of Saturated Water and Steam (Pressure) - I-P Units]

For standard dry air conditions, the enthalpies are [Properties of Saturated Water and Steam (Pressure) - I-P Units]

$$h_{f1} = 312.55 \text{ Btu/lbm}$$

$$h_{f2} = 180.18 \text{ Btu/lbm}$$

$$h_{\text{fg2}} = 970.07 \text{ Btu/lbm}$$

The percentage of flash steam produced is

Flash Steam

$$\% \text{ flash steam} = \frac{100(h_{f1} - h_{f2})}{h_{fg2}}$$

$$= \left(\frac{(100)\left(312.55 \, \dfrac{\text{Btu}}{\text{lbm}} - 180.18 \, \dfrac{\text{Btu}}{\text{lbm}} \right)}{970.07 \, \dfrac{\text{Btu}}{\text{lbm}}} \right)$$

$$= 13.6\%$$

The percentage of heat escaping in the flash steam is

$$\% \frac{\text{heat in flash steam}}{\text{heat in 105 psi water}} = \frac{(\% \text{ flash steam})(h_{fg2})}{h_{f1}}(100)$$

$$= \left(\frac{(0.136)\left(970.07 \, \dfrac{\text{Btu}}{\text{lbm}} \right)}{312.55 \, \dfrac{\text{Btu}}{\text{lbm}}} \right)(100)$$

$$= 42\%$$

The answer is (B).

Why Other Options Are Wrong

(A) This incorrect answer results when the quotient of h_{fg2} and h_{fg2} is used in the final calculation instead of the quotient of h_{fg2} and h_{f1}.

(C) This incorrect answer results when h_{fg1} and h_{fg2} are used in the final equation instead of h_{fg2} and h_{f1}.

(D) This incorrect answer results when h_{f2} is used in the final equation instead of h_{f1}.

SOLUTION 11

To find the Strouhal number, the Reynolds number must be calculated. The Reynolds number is defined as

Reynolds Number

$$\text{Re} = \frac{v_0 d}{\nu}$$

To calculate the Reynolds number, a current velocity is needed. From the answer options, a reasonable choice is 1.0 knots, which is 1.688 ft/sec. The kinematic viscosity of seawater at 50°F is 1.410×10^{-5} ft²/sec. [Thermal and Physical Properties of Ideal Gases (at Room Temperature)]

The Reynolds number is

$$\text{Re} = \frac{\left(1.688 \, \dfrac{\text{ft}}{\text{sec}} \right)(23 \text{ in})\left(\dfrac{1 \text{ ft}}{12 \text{ in}} \right)}{1.410 \times 10^{-5} \, \dfrac{\text{ft}^2}{\text{sec}}} \approx 2.29 \times 10^5$$

From the Strouhal number versus Reynolds number graph, the Strouhal number is approximately 0.2.

The velocities can now be calculated using the Strouhal number equation.

$$S = \frac{nd}{v_0}$$

$$v_1 = \frac{n_1 d}{S}$$

$$= \frac{\left(0.310 \, \dfrac{\text{rad}}{\text{sec}} \right)\left(\dfrac{1 \text{ cycle}}{2\pi \text{ rad}} \right)\left(\dfrac{1 \text{ Hz}}{1 \, \dfrac{\text{cycle}}{\text{sec}}} \right)}{0.2}$$

$$\times (23 \text{ in})\left(\dfrac{1 \text{ ft}}{12 \text{ in}} \right)\left(\dfrac{1 \text{ knot}}{1.688 \, \dfrac{\text{ft}}{\text{sec}}} \right)$$

$$= 0.28 \text{ knots}$$

$$v_2 = \frac{n_2 d}{S}$$

$$= \frac{\left(1.81 \, \dfrac{\text{rad}}{\text{sec}} \right)\left(\dfrac{1 \text{ cycle}}{2\pi \text{ rad}} \right)\left(\dfrac{1 \text{ Hz}}{1 \, \dfrac{\text{cycle}}{\text{sec}}} \right)}{0.2}$$

$$\times (23 \text{ in})\left(\dfrac{1 \text{ ft}}{12 \text{ in}} \right)\left(\dfrac{1 \text{ knot}}{1.688 \, \dfrac{\text{ft}}{\text{sec}}} \right)$$

$$= 0.64 \text{ knots}$$

The answer is (A).

Why Other Options Are Wrong

(B) This incorrect answer does not convert the velocity from ft/sec to knots.

(C) This incorrect answer does not convert $\omega_{1,2}$ from rad/sec to hertz.

(D) This incorrect answer does not convert the diameter from inches to feet.

SOLUTION 12

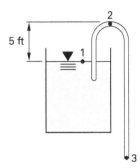

The total energy is constant, so the Bernoulli equation is

Bernoulli Equation

$$\frac{p_1}{\rho g} + z_1 + \frac{v_1^2}{2g} = \frac{p_2}{\rho g} + z_2 + \frac{v_2^2}{2g} + h_f$$

Adjusting for the need to convert pounds-mass to pounds-force, the equation becomes

$$\frac{p_1}{\rho} + \frac{v_1^2}{2g_c} + \frac{gz_1}{g_c} = \frac{p_2}{\rho} + \frac{v_2^2}{2g_c} + \frac{gz_2}{g_c} + \frac{gh_{losses}}{g_c}$$

From the law of conservation of mass for an incompressible fluid, the velocity at point 2 is the same as the velocity at point 3. The velocity at point 1 is negligible. Use point 1 as the zero-height reference. The Bernoulli equation simplifies to

$$\frac{p_1}{\rho} + 0 + 0 = \frac{p_2}{\rho} + \frac{v_2^2}{2g_c} + \frac{gz_2}{g_c} + 0$$

Solve for the pressure at point 2. The density of water at 70°F and 14.70 psia is 62.4 lbm/ft³. [Properties of Water at Standard Conditions]

$$p_2 = p_1 - \left(\frac{v_2^2}{2g_c} + \frac{gz_2}{g_c} \right)\rho$$

$$= p_1 - \left(\tfrac{1}{2}v_2^2 + gz_2 \right)\left(\frac{\rho}{g_c} \right)$$

$$= 14.70\ \frac{\text{lbf}}{\text{in}^2} - \left[\begin{array}{c} \left(\dfrac{1}{2}\right)\left(30\ \dfrac{\text{ft}}{\text{sec}}\right)^2 \\[6pt] + \left(32.17\ \dfrac{\text{ft}}{\text{sec}^2}\right)(5\ \text{ft}) \end{array} \right]$$

$$\times \frac{62.4\ \dfrac{\text{lbm}}{\text{ft}^3}}{\left(32.17\ \dfrac{\text{lbm-ft}}{\text{lbf-sec}^2}\right)\left(12\ \dfrac{\text{in}}{\text{ft}}\right)^2}$$

$$= 6.473\ \text{lbf/in}^2 \quad (6.5\ \text{psia})$$

The answer is (B).

Why Other Options Are Wrong

(A) This incorrect answer results from dividing the kinetic energy term by g_c instead of $2g_c$.

(C) This incorrect answer results from omitting the potential energy term.

(D) This incorrect answer results from omitting the kinetic energy term.

SOLUTION 13

The pipe can be drawn as shown.

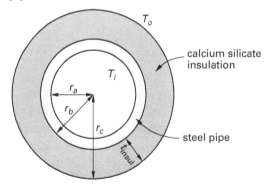

Use the equations shown to find the conduction and convection resistance of each layer.

Thermal Resistance (R)

$$R_{cond} = \frac{\ln\left(\dfrac{r_2}{r_1}\right)}{2\pi k L}$$

$$R_{conv} = \frac{1}{hA}$$

$$= \frac{1}{h(2\pi L r)}$$

$$R_{total} = \sum R$$

$$\dot{Q} = \frac{\Delta T}{R_{total}}$$

Combining the equations for the thermal resistance gives

$$R_{total} = R_a + R_{pipe} + R_b + R_{insul} + R_c$$

$$= \frac{1}{h_a(2\pi L r_a)} + \frac{\ln\dfrac{r_b}{r_a}}{2\pi k_{pipe} L} + \frac{1}{h_b(2\pi L r_b)}$$

$$+ \frac{\ln\dfrac{r_c}{r_b}}{2\pi k_{insul} L} + \frac{1}{h_c(2\pi L r_c)}$$

Solving for the overall heat transfer gives

$$\dot{Q} = \frac{2\pi L(T_i - T_o)}{\dfrac{1}{r_a h_a} + \dfrac{\ln\dfrac{r_b}{r_a}}{k_{pipe}} + \dfrac{1}{r_b h_b} + \dfrac{\ln\dfrac{r_c}{r_b}}{k_{insul}} + \dfrac{1}{r_c h_c}}$$

The temperature drop in the pipe wall is negligible, and all film coefficients are infinite, so the equation can be simplified further to solve for the heat loss through the pipe wall.

$$\dot{Q} = \frac{2\pi L(T_i - T_o)k_{insul}}{\ln\dfrac{r_c}{r_b}}$$

The inner diameter of 10 in schedule-80 steel pipe is 9.564 in, and the wall thickness is 0.593 in. [Schedule 80 Steel Pipe]

So, the outer radius of the pipe is

$$r_b = \frac{d_i}{2} + t_{pipe} = \frac{9.564 \text{ in}}{2} + 0.593 \text{ in}$$

$$= 5.375 \text{ in}$$

The equation for the outer radius of the insulation is

$$r_c = r_b + t_{insul}$$

Therefore,

$$\frac{r_c}{r_b} = \frac{r_b + t_{insul}}{r_b}$$

$$= 1 + \frac{t_{insul}}{r_b}$$

Substituting this into the larger equation and rearranging to solve for the thickness gives

$$\dot{Q} = \frac{2\pi L(T_i - T_o)k_{insul}}{\ln\left(1 + \dfrac{t_{insul}}{r_b}\right)}$$

$$\ln\left(1 + \frac{t_{insul}}{r_b}\right) = \frac{2\pi L(T_i - T_o)k_{insul}}{\dot{Q}}$$

$$1 + \frac{t_{insul}}{r_b} = \exp\left(\frac{2\pi L(T_i - T_o)k_{insul}}{\dot{Q}}\right)$$

$$t_{insul} = r_b\left(\exp\left(\frac{2\pi L(T_i - T_o)k_{insul}}{\dot{Q}}\right) - 1\right)$$

Find the heat loss rate needed so that the reduction in the water temperature is no more than 1°F. The specific heat capacity of water is 1 Btu/lbm-°F. [**Properties of Water at Standard Conditions**]

$$\dot{Q} = \dot{m}c_p\Delta T$$

$$= \left(7230\ \frac{\text{lbm}}{\text{hr}}\right)\left(1\ \frac{\text{Btu}}{\text{lbm-°F}}\right)(1\text{°F})$$

$$= 7230 \text{ Btu/hr}$$

The needed thickness is

$$t_{insul} = (5.375 \text{ in})\left(\exp\left(\frac{\begin{array}{c}(2\pi)(40 \text{ ft})\\ \times(400\text{°F} - 66\text{°F})\\ \times\left(0.046\ \dfrac{\text{Btu-ft}}{\text{hr-ft}^2\text{-°F}}\right)\end{array}}{7230\ \dfrac{\text{Btu}}{\text{hr}}}\right) - 1\right)$$

$$= 3.794 \text{ in} \quad (3.8 \text{ in})$$

The answer is (B).

Why Other Options Are Wrong

(A) This incorrect answer results when π is omitted from the calculation of the thickness.

(C) This incorrect answer results when the diameter is used to calculate the thickness instead of the radius.

(D) This incorrect answer results when 1 is not subtracted when calculating the thickness.

SOLUTION 14

The heat transfer on the inside of the handle is neglected, so the perimeter is the length of two short sides and one long side.

Modeling the pot handle as a fin, the equation for the heat transfer rate is

Fins

$$\dot{Q} = \sqrt{hPkA_c}\,(T_b - T_\infty)\tanh(mL_c)$$

Solve for each term of the heat transfer rate equation. From a table of metal properties, the heat conductivity of aluminum is 136 Btu/hr-ft-°R. [Properties of Metals - I-P Units]

$$P = w + 2t = 1 \text{ in} + (2)(0.5 \text{ in}) = 2 \text{ in}$$

$$A_c = (1 \text{ in})(0.5 \text{ in})$$

$$- \left(\begin{array}{c} \left(1 \text{ in} - \left(\dfrac{1}{16} \text{ in}\right)(2)\right) \\ \times \left(0.5 \text{ in} - \left(\dfrac{1}{16} \text{ in}\right)\right) \end{array} \right)$$

$$= 0.1172 \text{ in}^2$$

$$m = \sqrt{\frac{hP}{kA_c}} = \sqrt{\dfrac{1.2\,\dfrac{\text{Btu}}{\text{hr-ft}^2\text{-°F}}}{\left(136\,\dfrac{\text{Btu}}{\text{hr-ft-°R}}\right)\left(\dfrac{1°R}{1°F}\right)} \times \dfrac{(2 \text{ in})\left(\dfrac{1 \text{ ft}}{12 \text{ in}}\right)}{(0.12 \text{ in}^2)\left(\dfrac{1 \text{ ft}}{12 \text{ in}}\right)^2}}$$

$$= 1.344 \text{ ft}^{-1}$$

$$L_c = 7 \text{ in}$$

$$\sqrt{hPkA_c} = \sqrt{ \begin{array}{c} \left(1.2\,\dfrac{\text{Btu}}{\text{hr-ft}^2\text{-°F}}\right)(2 \text{ in})\left(\dfrac{1 \text{ ft}}{12 \text{ in}}\right) \\ \times \left(136\,\dfrac{\text{Btu}}{\text{hr-ft-°R}}\right)\left(\dfrac{1°F}{1°R}\right) \\ \times (0.1172 \text{ in}^2)\left(\dfrac{1 \text{ ft}}{12 \text{ in}}\right)^2 \end{array} }$$

$$= 0.1506 \text{ Btu/hr-°F}$$

Substituting everything back into the overall equation for the heat transfer rate, and assuming the heat transfer is from boiling water at 212°F to room temperature of 72°F,

$$\dot{Q} = \left(0.1506\,\dfrac{\text{Btu}}{\text{hr-°F}}\right)(212°F - 72°F)$$

$$\times \tanh\left((1.344 \text{ ft}^{-1})(7 \text{ in})\left(\dfrac{1 \text{ ft}}{12 \text{ in}}\right)\right)$$

$$= 13.81 \text{ Btu/hr} \quad (14 \text{ Btu/hr})$$

The answer is (A).

Why Other Options Are Wrong

(B) This incorrect answer results from adding 212°F to 72°F.

(C) This incorrect answer results from converting inches to feet instead of square inches to square feet in the square root term of the heat transfer equation.

(D) This incorrect answer results from not converting square inches to square feet in the square root term of the heat transfer equation.

SOLUTION 15

The needle valve in the line to the manifold controls the pressure to the manifold. The other needle valve allows oil to leave the line to the manifold and return to the pump intake line; therefore, this valve primarily controls the amount of oil flowing to the manifold.

The answer is (B).

Why Other Options Are Wrong

(A) This answer is incorrect because neither needle valve primarily controls the pressure at the pump intake.

(C) This answer is incorrect because neither needle valve primarily controls the flow rate to the pump intake.

(D) This answer is incorrect because neither needle valve primarily controls the flow rate to the pump intake.

SOLUTION 16

The acceleration of the block and vane can be found using Newton's second law.

Newton's Second Law (Equations of Motion)

$$\sum \mathbf{F} = ma$$

The mass of the block, m_{block}, is known, and the total force in the x-direction, F_x, is needed. There are two forces in the x-direction: one due to the water and the other, in the opposite direction, due to friction. Newton's second law for this situation can be restated as

$$F_{x,block} - F_{x,friction} = m_{block} a_{x,block}$$

Use the continuity equation to find the mass flow rate. The density of water at standard conditions is 62.4 lbm/ft³. [Properties of Water at Standard Conditions]

Continuity Equation

$$\dot{m} = \rho A v$$
$$= \left(62.4 \ \frac{lbm}{ft^3}\right)(0.005 \ ft^2)\left(150 \ \frac{ft}{sec}\right)$$
$$= 46.8 \ lbm/sec$$

The force that the block and vane exert on the water in the x-direction can be found by using the impulse-momentum principle.

Impulse-Momentum Principle

$$p_1 A_1 - p_2 A_2 \cos\alpha - F_x = Q\rho(v_2 \cos\alpha - v_1)$$
$$= \dot{m}(v_2 \cos\alpha - v_1)$$

$$F_x = -\dot{m}(v_2 \cos\alpha - v_1) + p_1 A_1 - p_2 A_2 \cos\alpha$$

The velocities approaching and leaving the vane are the same. Points 1 and 2 are at atmospheric pressure, so p_1 and p_2 are 0 psig, and the equation simplifies to

$$F_{x,water} = \frac{-\dot{m}(v_2\cos\alpha - v_1)}{g_c}$$

$$= \frac{-\left(46.8 \ \dfrac{lbm}{sec}\right)}{32.17 \ \dfrac{lbm\text{-}ft}{lbf\text{-}sec^2}}$$
$$\times\left(\left(150 \ \dfrac{ft}{sec}\right)\cos30° - 150 \ \dfrac{ft}{sec}\right)$$

$$= 29.21 \ lbf$$

The force that the water exerts on the block and vane is equal in magnitude and opposite in direction, so the x-component of the force, $F_{x,block}$, is 29.21 lbf to the right.

For the friction force, the y-force due to the water is needed. The force that the block and vane exert on the water in the y-direction can be found by using the impulse-momentum principle.

Impulse-Momentum Principle

$$F_y - W - p_2 A_2 \sin\alpha = Q\rho(v_2 \sin\alpha - 0)$$
$$= \dot{m}(v_2 \sin\alpha - 0)$$

$$F_y = \dot{m}(v_2 \sin\alpha) + W + p_2 A_2 \sin\alpha$$

The weight of the water, W, inside the control volume is negligible, and p_2 is zero as before, so this equation simplifies to

$$F_{y,water} = \frac{\dot{m}(v_2 \sin\alpha)}{g_c}$$

$$= \frac{\left(46.8 \ \dfrac{lbm}{sec}\right)\left(\left(150 \ \dfrac{ft}{sec}\right)\sin 30°\right)}{32.17 \ \dfrac{lbm\text{-}ft}{lbf\text{-}sec^2}}$$

$$= 109.0 \ lbf$$

The force that the water exerts on the block and vane is equal in magnitude and opposite in direction. So $F_{y,block}$ = 109.0 lbf down. Sum the forces in the y-direction, and solve for the normal force.

$$N - \frac{W_{block}g}{g_c} - F_{y,water} = 0$$

$$N = \frac{W_{block}g}{g_c} + F_{y,water}$$

$$= \frac{(50 \ lbm)\left(32.17 \ \dfrac{ft}{sec^2}\right)}{32.17 \ \dfrac{lbm\text{-}ft}{lbf\text{-}sec^2}} + 109.0 \ lbf$$

$$= 159.0 \ lbf$$

The friction force in the x-direction is

Friction

$$F_{x,friction} = \mu_s N$$
$$= (0.1)(159.0 \ lbf)$$
$$= 15.90 \ lbf$$

The mass is in pounds-mass, so conversion to pounds-force using the gravitational constant is necessary.

$$F_{x,block} - F_{x,friction} = \left(\frac{m_{block}}{g_c}\right) a_{x,block}$$

$$a_{x,block} = \frac{(F_{x,block} - F_{x,friction})g_c}{m_{block}}$$

$$= \frac{(29.21 \ lbf - 15.90 \ lbf)\left(32.17 \ \dfrac{lbm\text{-}ft}{lbf\text{-}sec^2}\right)}{50 \ lbm}$$

$$= 8.572 \ ft/sec^2 \quad (8.6 \ ft/sec^2)$$

The answer is (B).

Why Other Options Are Wrong

(A) This incorrect answer results when the gravitational constant, g_c, is neglected.

(C) This incorrect answer results when friction is neglected.

(D) This incorrect answer results when the friction force is added to the force of the water on the block.

SOLUTION 17

The flow is isentropic up to and after the normal shock. Using the normal shock relationships, the equation for the Mach number just after the shock wave is

Normal Shock Relationships

$$M_2 = \sqrt{\frac{(k-1)M_1^2 + 2}{2k\,M_1^2 - (k-1)}}$$

For air, the ratio of specific heats, k, is $= 1.40$. [Thermal and Physical Properties of Ideal Gases (at Room Temperature)]

The Mach number just after the shock wave is

$$M_2 = \sqrt{\frac{(1.40-1)(1.8)^2 + 2}{(2)(1.40)(1.8)^2 - (1.40-1)}} = 0.62$$

Since the shock stands at the entrance, the entrance area is the throat area, A^*. The Mach number at the exit can be found using the equation shown and the isentropic flow table.

$$\frac{A_{\text{exit}}}{A^*}$$

Determine the ratio using known ratios.

$$\frac{A_{\text{exit}}}{A_{\text{in}}} = \left(\frac{A_{\text{exit}}}{A^*}\right)_{M_{\text{exit}}} \left(\frac{A^*}{A_{\text{in}}}\right)_{M_2 = 0.62}$$

$$3 = \left(\frac{A_{\text{exit}}}{A^*}\right)_{M_{\text{exit}}} \left(\frac{A^*}{A_{\text{in}}}\right)_{M_2 = 0.62}$$

$$3\left(\frac{A_{\text{in}}}{A^*}\right)_{M_2 = 0.62} = \left(\frac{A_{\text{exit}}}{A^*}\right)_{M_{\text{exit}}}$$

From a compressible-flow functions table, A_{in}/A^* for a Mach number of 0.62 is 1.1656. [One-Dimensional Isentropic Compressible-Flow Functions $k = 1.4$]

Thus,

$$(3)(1.1656) = \left(\frac{A_{\text{exit}}}{A^*}\right)_{M_{\text{exit}}}$$

$$\left(\frac{A_{\text{exit}}}{A^*}\right)_{M_{\text{exit}}} = 3.5$$

Interpolating from a table of isentropic compressible-flow functions, this ratio corresponds to a Mach number of approximately 0.17. [One-Dimensional Isentropic Compressible-Flow Functions $k = 1.4$]

The answer is (A).

Why Other Options Are Wrong

(B) This incorrect answer results when the p/p_0 ratio at $\text{Ma} = 0.62$ is used.

(C) This incorrect answer results when the Mach number is divided by the exit-to-throat area.

(D) This incorrect answer results when the supersonic solution is found.

SOLUTION 18

From the steam tables, using interpolation, the specific volume of the steam is 1.0825 ft³/lbm. [Properties of Superheated Steam - I-P Units]

A 6 in schedule-40 pipe has an internal diameter and internal area of [Schedule 40 Steel Pipe]

$$d_i = 6.065 \text{ in}$$

$$A = \pi\left(\frac{d_i}{2}\right)^2 = \pi\left(\frac{6.065 \text{ in}}{2}\right)^2 = 28.89 \text{ in}^2$$

The equation for the steam velocity is

Continuity Equation

$$\dot{m} = \rho Q = \rho A \text{v}$$

$$\text{v} = \frac{Q}{A} = \frac{\dot{m}}{\rho A} = \frac{\dot{m}v}{A}$$

Therefore,

$$v = \frac{\dot{m}v}{A}$$

$$= \frac{\left(50{,}000 \ \frac{\text{lbm}}{\text{hr}}\right)\left(\frac{1 \ \text{hr}}{3600 \ \text{sec}}\right)\left(1.0825 \ \frac{\text{ft}^3}{\text{lbm}}\right)}{(28.89 \ \text{in}^2)\left(\frac{1 \ \text{ft}}{12 \ \text{in}}\right)^2}$$

$$= 75 \ \text{ft/sec}$$

The head loss due to friction is

Head Loss Due to Flow: Darcy-Weisbach Equation

$$h_f = f\frac{Lv^2}{2Dg}$$

$$= (0.02)\left(\frac{(100 \ \text{ft})\left(75 \ \frac{\text{ft}}{\text{sec}}\right)^2}{(2)(6.065 \ \text{in})\left(\frac{1 \ \text{ft}}{12 \ \text{in}}\right)\left(32.17 \ \frac{\text{ft}}{\text{sec}^2}\right)}\right)$$

$$= 346 \ \text{ft}$$

The pressure drop is

Bernoulli Equation

$$p_1 - p_2 = \rho g h_f$$

$$p_1 - p_2 = \rho\left(\frac{g}{g_c}\right)h_f$$

$$\Delta p = \left(\frac{h_f}{v}\right)\left(\frac{g}{g_c}\right)$$

$$= \left(\frac{(346 \ \text{ft})\left(12 \ \frac{\text{in}}{\text{ft}}\right)}{\left(1.0825 \ \frac{\text{ft}^3}{\text{lbm}}\right)\left(12 \ \frac{\text{in}}{\text{ft}}\right)^3}\right)$$

$$\times \left(\frac{32.17 \ \frac{\text{ft}}{\text{sec}^2}}{32.17 \ \frac{\text{ft-lbm}}{\text{sec}^2\text{-lbf}}}\right)$$

$$= 2.2 \ \text{lbf/in}^2$$

The answer is (A).

Why Other Options Are Wrong

(B) This incorrect answer results when the wrong internal pipe diameter is read from the schedule-40 table.

(C) This incorrect answer results when a specific volume of 1.825 ft^3/lbm, instead of 1.0825 ft^3/lbm, is used in the velocity calculation.

(D) This incorrect answer results when the conversion factor for v is squared instead of cubed.

SOLUTION 19

The straight-line depreciation for each year is

Depreciation: Straight Line

$$D_f = \frac{C - S_n}{n}$$

$$= \frac{\$130{,}000 - \$10{,}000}{4}$$

$$= \$30{,}000$$

Taxes are owed on net profits less depreciation, so

$$T = r_{\text{tax}}(P - D) = (0.45)(\$45{,}000 - \$30{,}000) = \$6750$$

Then, the net income for each year is

$$\text{NI} = P - T = \$45{,}000 - \$6750 = \$38{,}250$$

The present worth of the equipment is the sum of the present worths of the net annual income and the salvage value minus the cost. From economic factor tables, given an interest rate of 8% and a time period of 4 years, the cash flow factor to convert an annual cost to a present worth is 3.3121, and the cash flow factor to convert a future worth to a present worth is 0.7350. [Economic Factor Tables]

$$\begin{aligned} P_{\text{income}} &= (\text{NI})(P/A, \, i, \, n) \\ &= (\$38{,}250)(P/A, \, 8\%, \, 4) \\ &= (\$38{,}250)(3.3121) \\ &= \$126{,}688 \\ P_{\text{salvage}} &= S(P/F, \, i, \, n) \\ &= (\$10{,}000)(P/F, \, 8\%, \, 4) \\ &= (\$10{,}000)(0.7350) \\ &= \$7350 \\ P &= P_{\text{income}} + P_{\text{salvage}} - C \\ &= \$126{,}688 + \$7350 - \$130{,}000 \\ &= \$4038 \ (\$4000) \end{aligned}$$

The answer is (C).

Why Other Options Are Wrong

(A) This incorrect answer results when the future value is calculated instead of the present value and the signs are reversed.

(B) This incorrect answer results when the signs are reversed.

(D) This incorrect answer results when the future value is calculated instead of the present value.

SOLUTION 20

The equation for the power output of the turbine is

Turbines

$$\dot{W} = \dot{m}(h_i - h_e)$$

Using the superheated steam tables, at $p_1 = 300$ psia and $T_1 = 740°F$, $h_i = 1389.7$ Btu/lbm. [Properties of Superheated Steam - I-P Units]

Using the saturated water and steam tables, at $p_2 = 2$ psia, $h_f = 94.00$ Btu/lbm and $h_{fg} = 1021.74$ Btu/lbm. [Properties of Saturated Water and Steam (Pressure) - I-P Units]

The enthalpy at the exit is

Properties for Two-Phase (Vapor-Liquid) Systems

$$h_e = h_f + xh_{fg}$$
$$= 94.00 \frac{Btu}{lbm} + (0.90)\left(1021.74 \frac{Btu}{lbm}\right)$$
$$= 1013.6 \text{ Btu/lbm}$$

Rearranging the equation for the power output of the turbine, the mass flow rate is

$$\dot{m} = \frac{\dot{W}}{(h_i - h_e)}$$
$$= \frac{\left(\dfrac{(6700\,\text{hp})\left(\dfrac{1\,\dfrac{Btu}{hr}}{3.93 \times 10^{-4}\,\text{hp}}\right)}{1389.7\,\dfrac{Btu}{lbm} - 1013.6\,\dfrac{Btu}{lbm}}\right)}{3600\,\dfrac{sec}{hr}}$$
$$= 12.6 \text{ lbm/sec} \quad (13 \text{ lbm/sec})$$

The answer is (B).

Why Other Options Are Wrong

(A) This incorrect answer results when h_{fg} is used for h_2.

(C) This incorrect answer results when the steam quality is omitted when calculating the steam enthalpy.

(D) This incorrect answer results when the conversion 1 hr/60 sec is used.

SOLUTION 21

The net positive suction head available (NPSHA) is defined as

Centrifugal Pump Characteristics

$$\text{NPSH}_A = h_p + h_z - h_{vpa} - h_f$$

Item I will increase the NPSHA because a larger diameter pipe will reduce the fluid velocity at the pump inlet and the reduced fluid velocity will reduce the friction loss, h_f, as well.

Item II will *not* increase the NPSHA because throttling the input will increase the fluid velocity. This will increase the increase the friction head loss and decrease the NPSHA.

Item III will increase the NPSHA because pressurizing the supply tank will increase the atmospheric pressure, h_p.

Item IV will increase the NPSHA because increasing the height above the pump increases the static suction head, h_z.

The answer is (D).

Why Other Options Are Wrong

(A) This answer is wrong because item III will also increase the NPSHA.

(B) This answer is wrong because item II will not increase the NPSHA.

(C) This answer is wrong because item II will not increase the NPSHA.

SOLUTION 22

Item I is a pipe fitting classification and is an abbreviation for American National Standard Taper pipe thread. NPT is also used as an abbreviation for the following:

American Standard Pipe Tapered Thread
National Pipe Thread
National Pipe Tapered

Item II is a pipe fitting classification and is also an abbreviation for the Society of Automotive Engineers.

Item III is a screw thread classification and is an abbreviation for Unified Coarse.

Item IV is the abbreviation for the American Society of Heating, Refrigerating and Air-Conditioning Engineers.

The answer is (A).

Why Other Options Are Wrong

(B) This answer is wrong because item III is a screw thread classification.

(C) This answer is wrong because item III is a screw thread classification.

(D) This answer is wrong because item III is a screw thread classification and item IV is the abbreviation for the American Society of Heating, Refrigeration, and Air-Conditioning Engineers.

SOLUTION 23

The heat transfer rate needed to obtain the required steam production is

$$q = \dot{m}\Delta h_{\text{water}}$$
$$= \dot{m}(h_{\text{steam}} - h_{\text{water}})$$

From steam tables, find the enthalpy of the saturated water entering the drum boiler at 580°F. [Properties of Saturated Water and Steam (Temperature) - I-P Units]

$$h_{\text{water}} = h_{f,580°F} = 589.05 \text{ Btu/lbm}$$

From steam tables, find the enthalpy for saturated water and the heat of vaporization at 1400 psia. [Properties of Saturated Water and Steam (Pressure) - I-P Units]

$$h_{f,1400\,\text{psia}} = 598.85 \text{ Btu/lbm}$$
$$h_{\text{fg},1400\,\text{psia}} = 575.60 \text{ Btu/lbm}$$

Calculate the enthalpy of steam at 1400 psia and 95% quality.

Properties for Two-Phase (Vapor-Liquid) Systems

$$h_{\text{steam}} = h_f + x h_{\text{fg}}$$
$$= 598.85 \frac{\text{Btu}}{\text{lbm}} + (0.95)\left(575.60 \frac{\text{Btu}}{\text{lbm}}\right)$$
$$= 1145.67 \text{ Btu/lbm}$$

Only 92% of the transferred heat is used, so the total heat transferred from the flue gas must be

$$q_{\text{total}} = \frac{\dot{m}(h_{\text{steam}} - h_{\text{water}})}{\eta}$$
$$= \frac{\left(540,000 \frac{\text{lbm}}{\text{hr}}\right)\left(1145.67 \frac{\text{Btu}}{\text{lbm}} - 589.05 \frac{\text{Btu}}{\text{lbm}}\right)}{0.92}$$
$$= 326,711,739 \text{ Btu/hr}$$

The heat transfer rate as a function of the temperature difference is given by

$$q_{\text{total}} = UA(T_1 - T_2)$$

Solving for the heat-exchange area,

$$A = \frac{q_{\text{total}}}{U(T_1 - T_2)}$$
$$= \frac{326,711,739 \frac{\text{Btu}}{\text{hr}}}{\left(218.2 \frac{\text{Btu}}{\text{hr-ft}^2\text{-°F}}\right)(32°F)}$$
$$= 46,790.75 \text{ ft}^2$$

The total number of tubes required is

$$n_{\text{tubes}} = \frac{A}{A_{\text{tube}}}$$
$$= \frac{46,790.75 \text{ ft}^2}{5.3 \frac{\text{ft}^2}{\text{tube}}}$$
$$= 8828 \text{ tubes} \quad (8800 \text{ tubes})$$

The answer is (B).

Why Other Options Are Wrong

(A) This incorrect answer results when the total heat transfer is multiplied by the efficiency instead of divided by it.

(C) This incorrect answer results when the steam quality is omitted when calculating the steam enthalpy.

(D) This incorrect answer results when the temperature difference is neglected in the calculation of the required area.

SOLUTION 24

The suction lift can be calculated using the net positive suction head available (NPSHA) equation.

Centrifugal Pump Characteristics
$$\text{NPSH}_A = h_p + h_z - h_{\text{vpa}} - h_f$$

From a table of water properties, the specific weight of water is 62.4 lbf/ft³. [Properties of Water at Standard Conditions]

The atmospheric head is

$$h_p = \frac{p_p}{\rho g} = \frac{p_p}{\gamma} = \frac{\left(14.7 \, \frac{\text{lbf}}{\text{in}^2}\right)\left(12 \, \frac{\text{in}}{\text{ft}}\right)^2}{62.4 \, \frac{\text{lbf}}{\text{ft}^3}}$$

$$= 34 \text{ ft}$$

Also from a table of water properties, the vapor pressure of water at 70°F is 0.36 psi. [Properties of Water (I-P Units)]

The vapor pressure head is

$$h_{\text{vpa}} = \frac{p_{\text{vpa}}}{\rho g} = \frac{p_{\text{vpa}}}{\gamma} = \frac{\left(0.36 \, \frac{\text{lbf}}{\text{in}^2}\right)\left(12 \, \frac{\text{in}}{\text{ft}}\right)^2}{62.4 \, \frac{\text{lbf}}{\text{ft}^3}}$$

$$= 0.83 \text{ ft}$$

The friction loss head is 5 ft. The static suction head can now be calculated.

$$\text{NPSH}_A = h_p + h_z - h_{\text{vpa}} - h_f$$
$$h_z = \text{NPSH}_A - h_p + h_{\text{vpa}} + h_f$$
$$= 10 \text{ ft} - 34 \text{ ft} + 0.83 \text{ ft} + 5 \text{ ft}$$
$$= -18.17 \text{ ft} \quad (18 \text{ ft})$$

The negative sign indicates that the pump is above the free level of the supply surface, and there is 18 ft of suction lift.

The answer is (B).

Why Other Options Are Wrong

(A) This answer results when the net positive suction head and the suction lift are incorrectly thought to be equal.

(C) This incorrect answer results when the atmospheric head is taken as 43 ft instead of 34 ft.

(D) This incorrect answer results when the friction loss head and vapor pressure head are added to the right side of the NPSHA equation instead of subtracted.

SOLUTION 25

From the steam tables, at 212°F, [Properties of Saturated Water and Steam (Temperature) - I-P Units]

$$h_{\text{fg}} = 970.09 \text{ Btu/lbm}$$

The equivalent to one boiler horsepower in Btu/hr is

$$P = \dot{m}\Delta h = \left(34.5 \, \frac{\text{lbm}}{\text{hr}}\right)\left(970.09 \, \frac{\text{Btu}}{\text{lbm}}\right)$$
$$= 33{,}468 \text{ Btu/hr} \quad (34{,}000 \text{ Btu/hr})$$

The answer is (A).

Why Other Options Are Wrong

(B) This incorrect answer results from using h_g at 212°F.

(C) This incorrect answer results from transposing the 3 and 4 in the mass flow term.

(D) This incorrect answer results from using h_{fg} in SI units at 100°C.

SOLUTION 26

The set pressure that will yield the same safety valve pressure can be found by equating the two pressure equations.

$$1.03(\text{set pressure}) + 14.7 \, \frac{\text{lbf}}{\text{in}^2} = \text{set pressure}$$
$$+ 2 \, \frac{\text{lbf}}{\text{in}^2} + 14.7 \, \frac{\text{lbf}}{\text{in}^2}$$

$$\text{set pressure} = 67 \text{ lbf/in}^2$$

Using an arbitrary set pressure of 70 psi as a test case,

$$p_{\text{first},70} = (1.03)\left(70 \, \frac{\text{lbf}}{\text{in}^2}\right) + 14.7 \, \frac{\text{lbf}}{\text{in}^2} = 86.8 \, \frac{\text{lbf}}{\text{in}^2}$$
$$p_{\text{second},70} = 70 \, \frac{\text{lbf}}{\text{in}^2} + 2 \, \frac{\text{lbf}}{\text{in}^2} + 14.7 \, \frac{\text{lbf}}{\text{in}^2} = 86.7 \, \frac{\text{lbf}}{\text{in}^2}$$

Because both equations are linear, the first equation will always yield a larger safety valve for any set pressure above 67 psi.

The answer is (B).

Why Other Options Are Wrong

(A) This incorrect answer results from subtracting 14.7 from 67.

(C) This incorrect answer results from adding 14.7 to 67.

(D) This incorrect answer results from adding (2)(14.7) to 67.

SOLUTION 27

Find the required thickness.

$$t_{req} = \frac{pr}{SE - 0.6p}$$

$$= \frac{\left(250 \frac{\text{lbf}}{\text{in}^2}\right)(34 \text{ in})}{\left(17{,}100 \frac{\text{lbf}}{\text{in}^2}\right)(1.0) - (0.6)\left(250 \frac{\text{lbf}}{\text{in}^2}\right)}$$

$$= 0.501 \text{ in}$$

The required reinforcement area is

$$A_{req} = Dt_{req}$$
$$= (4.5 \text{ in})(0.501 \text{ in})$$
$$= 2.25 \text{ in}^2$$

The area available in the shell is

$$A_s = D(t - t_{req}) = (4.5 \text{ in})(0.625 \text{ in} - 0.501 \text{ in})$$

$$= 0.558 \text{ in}^2$$
$$A_s + A_n = 0.558 \text{ in}^2 + 0.518 \text{ in}^2 = 1.08 \text{ in}^2$$

Since this is less than the required area, the amount of additional reinforcement needed is

$$A_{add} = 2.25 \text{ in}^2 - 1.08 \text{ in}^2$$
$$= 1.17 \text{ in}^2 \quad (1.2 \text{ in}^2)$$

The answer is (B).

Why Other Options Are Wrong

(A) This incorrect answer results when the wrong diameter for the nozzle is used.

(C) This incorrect answer results when an incorrect efficiency of 0.6 is used to calculate t_{req}.

(D) This incorrect answer results when the vessel's diameter is used to calculate t_{req}.

SOLUTION 28

Find the outside diameter of the pipe. From a schedule-40 steel pipe properties table, the inside diameter of a 3 in schedule-40 steel pipe is 3.068 in, and the wall thickness of a 3 in schedule-40 steel pipe is 0.216 in. [Schedule 40 Steel Pipe]

$$D_o = D_i + 2(\text{wall thickness})$$
$$= 3.068 \text{ in} + (2)(0.216 \text{ in})$$
$$= 3.5 \text{ in}$$

The minimum thickness is

$$t_{min} = \frac{pD_o}{2SE}$$

$$= \frac{\left(250 \frac{\text{lbf}}{\text{in}^2}\right)(3.5 \text{ in})}{(2)\left(17{,}100 \frac{\text{lbf}}{\text{in}^2}\right)(1)}$$

$$= 0.02558 \text{ in}$$

The remaining corrosion allowance (RCA) is

$$RCA = t_{actual} - t_{min}$$
$$= 0.175 \text{ in} - 0.02558 \text{ in}$$
$$= 0.14942 \text{ in} \quad (0.15 \text{ in})$$

The answer is (B).

Why Other Options Are Wrong

(A) This incorrect answer results from neglecting the 2 in the denominator.

(C) This incorrect answer results from using a wrong diameter of 3 in, a wrong stress value of 20,000 lbf/in², and a factor of safety of 1.67, which is not needed because ASME Sec. II gives the corrected stress value.

(D) This incorrect answer results from using the original pipe thickness for the actual thickness.

SOLUTION 29

In the equation for f, N is the rotational speed in revolutions per second, rps; P is the load per unit of projected bearing area in psi; r is the journal radius in inches; c is the radial clearance in inches; and f is a unitless value. [Journal Bearing Design]

Perform a unit analysis.

$$f = \left(\pi^2 \frac{\text{rad}}{\text{rev}}\right)\mu \frac{\left(\frac{\text{rev}}{\text{sec}}\right)}{\left(\frac{\text{lbf}}{\text{in}^2}\right)}\left(\frac{\text{in}}{\text{in}}\right)$$

$$= \pi^2 \mu \left(\frac{\text{rad-in}^2}{\text{lbf-sec}}\right)$$

The dynamic viscosity must have units of

$$\frac{\text{lbf-sec}}{\text{in}^2} = \text{reyn}$$

The answer is (D).

Why Other Options Are Wrong

(A) This incorrect answer is in SI units.

(B) This incorrect answer is in pounds mass.

(C) This incorrect answer uses SI units for dynamic viscosity.

SOLUTION 30

The compression ratio of the compressor in a refrigeration cycle is the ratio of the discharge pressure to the suction pressure. The pressures must be absolute. Using the R134a table, at 120°F the pressure is 185.860 psia, and at 0°F the pressure is 21.171 psia. [**Refrigerant 134a (1,1,1,2-Tetrafluoroethane) Properties of Saturated Liquid and Saturated Vapor**]

The compression ratio is

$$r_p = \frac{p_{\text{discharge}}}{p_{\text{suction}}} = \frac{185.860 \dfrac{\text{lbf}}{\text{in}^2}}{21.171 \dfrac{\text{lbf}}{\text{in}^2}} = 8.8$$

The answer is (C).

Why Other Options Are Wrong

(A) This incorrect answer results from dividing p_{suction} by $p_{\text{discharge}}$.

(B) This incorrect answer results from using gauge pressures.

(D) This incorrect answer results from using R123.

SOLUTION 31

The total enthalpy in the steam, h, equals 800 Btu/lbm plus the initial enthalpy in the feedwater. The initial enthalpy of the feedwater at 280°F is 249.20 Btu/lbm. [**Properties of Saturated Water and Steam (Temperature) - I-P Units**]

The total enthalpy, then, is

$$\begin{aligned} h_{\text{total}} &= h_{\text{added}} + h_{\text{feed}} \\ &= 800 \ \frac{\text{Btu}}{\text{lbm}} + 249.20 \ \frac{\text{Btu}}{\text{lbm}} \\ &= 1049 \ \text{Btu/lbm} \end{aligned}$$

From the steam tables, at 1200 psia, [**Properties of Saturated Water and Steam (Pressure) - I-P Units**]

$$h_f = 571.94 \ \text{Btu/lbm}$$
$$h_{\text{fg}} = 612.29 \ \text{Btu/lbm}$$

The steam quality is

Properties for Two-Phase (Vapor-Liquid) Systems

$$h = h_f + x h_{\text{fg}}$$

$$1049 \ \frac{\text{Btu}}{\text{lbm}} = 571.94 \ \frac{\text{Btu}}{\text{lbm}} + x\left(612.29 \ \frac{\text{Btu}}{\text{lbm}}\right)$$

$$x = 0.779 \quad (78\%)$$

The answer is (C).

Why Other Options Are Wrong

(A) This incorrect answer results when 800 Btu/lbm is used for the total energy.

(B) This incorrect answer results when 249.20 Btu/lbm is used for the total enthalpy.

(D) This incorrect answer results when 1094 is entered into the calculator instead of 1049.

SOLUTION 32

Statement I is false because, as the hint suggests, isothermal compression requires less power, so heat transfer occurs and the process is not isentropic.

Statement II is also false. An isentropic process is both adiabatic and reversible, but an adiabatic process need not be reversible. So although all isentropic processes are adiabatic, not all adiabatic processes are isentropic.

The absolute temperature is 70°F + 460° = 530°R. [**Temperature Conversions**]. For this absolute temperature, the specific heat, c_p, is approximately 0.24 Btu/lbm-°R, and the ratio of specific heats is approximately 1.4. [**Properties of Air at Low Pressure, per Pound**]

The final temperature can be calculated from the initial temperature using the equation shown.

Isentropic Flow Relationships

$$\frac{p_2}{p_1} = \left(\frac{T_2}{T_1}\right)^{\frac{k}{k-1}}$$

$$\begin{aligned} T_2 &= T_1\left(\frac{p_2}{p_1}\right)^{\frac{k-1}{k}} \\ &= (530°\text{R})\left(\frac{200 \ \dfrac{\text{lbf}}{\text{in}^2}}{14.7 \ \dfrac{\text{lbf}}{\text{in}^2}}\right)^{\frac{1.4-1}{1.4}} \\ &= 1120°\text{R} \end{aligned}$$

The power needed is

Compressors

$$\dot{W} = \dot{m}c_p(T_2 - T_1)$$

$$= \left(1\ \frac{\text{lbm}}{\text{sec}}\right)\left(\frac{3600\ \text{sec}}{1\ \text{hr}}\right)\left(0.24\ \frac{\text{Btu}}{\text{lbm-°R}}\right)$$

$$\times (1120°\text{R} - 530°\text{R})\left(\frac{1\ \text{hp}}{2544\ \dfrac{\text{Btu}}{\text{hr}}}\right)$$

$$= 200\ \text{hp}$$

Thus, only statement III is true.

The answer is (C).

Why Other Options Are Wrong

(A) This answer is incorrect because statement I is false as explained above.

(B) This answer is incorrect because statement II is false as explained above.

(D) This answer is incorrect because statement I is false as explained above.

SOLUTION 33

Using the given equation,

$$t_{\text{min}} = D\sqrt{\frac{Cp}{SE}}$$

$$= (3\ \text{ft})\left(12\ \frac{\text{in}}{\text{ft}}\right)\sqrt{\frac{(0.20)\left(150\ \dfrac{\text{lbf}}{\text{in}^2}\right)}{\left(17{,}500\ \dfrac{\text{lbf}}{\text{in}^2}\right)(1)}}$$

$$= 1.5\ \text{in}$$

The thickness of the head with the rate of corrosion taken into account is

$$t_{\text{head}} = t_{\text{min}} + (\text{CR})L$$

$$= 1.5\ \text{in} + \left(\frac{8\ \dfrac{\text{mil}}{\text{yr}}}{1000\ \dfrac{\text{mil}}{\text{yr}}}\right)(30\ \text{yr})$$

$$= 1.74\ \text{in}\quad (1.7\ \text{in})$$

The answer is (C).

Why Other Options Are Wrong

(A) The minimum design thickness must be greater than just the corrosion amount of 0.24 in over 30 years or the head will not be able to withstand the pressure.

(B) This incorrect answer results when the diameter is not converted to inches.

(D) This incorrect answer results when the attachment factor is ignored.

SOLUTION 34

Water from the air will condense onto the inner window surfaces (fogging) when the temperature of the inner surface of the window equals the dew-point temperature of the air. The dew-point temperature of the air at 68°F and 75% relative humidity can be found using a psychrometric chart. To use the psychrometric chart, first locate the dry-bulb temperature of 68°F on the bottom horizontal axis of the chart. Then follow this line straight upward to the 75% relative humidity line. From this point, go horizontally to the left to the curved saturation line. The dew-point temperature can be read as approximately 60°F. This temperature can also be read by dropping straight back down to the horizontal axis. [ASHRAE Psychrometric Chart No. 1 - Normal Temperature at Sea Level]

The answer is (D).

Why Other Options Are Wrong

(A) This incorrect answer results when the enthalpy value is used.

(B) This incorrect answer results when a horizontal line to the right is used and an enthalpy value is read.

(C) This incorrect answer results when the 58°F dry-bulb temperature line is used.

SOLUTION 35

The drag force on the carrier is calculated using the equation shown.

Drag Force

$$F_D = \frac{C_D \rho \text{v}^2 A}{2g_c}$$

The factor g_c is needed because the density is in units of lbm/ft³. The density, ρ, at 68°F is 0.0752 lbm/ft³. [Properties of Air at Atmospheric Pressure]

The drag force is

$$F_D = \frac{C_D \rho \mathrm{v}^2 A}{2g_c}$$

$$= \frac{(1.25)\left(0.0752 \; \dfrac{\text{lbm}}{\text{ft}^3}\right)}{\left[\left(50 \; \dfrac{\text{mi}}{\text{hr}} + 10 \; \dfrac{\text{mi}}{\text{hr}}\right)\left(\dfrac{5280 \; \text{ft}}{1 \; \text{mi}}\right)\left(\dfrac{1 \; \text{hr}}{3600 \; \text{sec}}\right)\right]^2}{(2)\left(32.17 \; \dfrac{\text{ft-lbm}}{\text{lbf-sec}^2}\right)}$$

$$= 37.7 \; \text{lbf}$$

Relative velocity must be used because 60 mph is what the cargo carrier "feels" in drag force speed.

The additional power needed is

Power and Efficiency

$$P = \mathbf{F} \cdot \mathbf{v}$$

$$= (37.7 \; \text{lbf})\left(50 \; \frac{\text{mi}}{\text{hr}} + 10 \; \frac{\text{mi}}{\text{hr}}\right)\left(5280 \; \frac{\text{ft}}{\text{mi}}\right)$$

$$\times \frac{\left(\dfrac{1 \; \text{hp}}{550 \; \dfrac{\text{ft-lbf}}{\text{sec}}}\right)}{3600 \; \dfrac{\text{sec}}{\text{hr}}}$$

$$= 6.0 \; \text{hp}$$

The answer is (C).

Why Other Options Are Wrong

(A) This incorrect answer results when 60 mph is not converted to ft/sec.

(B) This incorrect answer results when 50 mph is used, not 60 mph.

(D) This incorrect answer results when the two in the denominator is neglected.

SOLUTION 36

When compressed, the air is reduced to 1/16 of its original volume, but the mass of air remains the same. Therefore, the density of the gas is increased by a factor of 16. The ratio of specific heats, k, for air is 1.40, and the initial pressure, p_1, is atmospheric at 14.7 psi. [Thermal and Physical Properties of Ideal Gases (at Room Temperature)]

For isentropic compression, the flow relationship is

Isentropic Flow Relationships

$$\frac{p_2}{p_1} = \left(\frac{\rho_2}{\rho_1}\right)^k$$

The subscripts 1 and 2 represent the initial and final states, respectively. Solving for p_2, the air pressure at the end of the compression process is

$$p_2 = p_1 \left(\frac{\rho_2}{\rho_1}\right)^k$$

$$= \left(14.7 \; \frac{\text{lbf}}{\text{in}^2}\right)\left(\frac{16\rho_1}{\rho_1}\right)^{1.40}$$

$$= 713 \; \text{lbf/in}^2 \quad (710 \; \text{psi})$$

The answer is (D).

Why Other Options Are Wrong

(A) This incorrect answer results when the 1.4th root of 16 is taken instead of the 1.4th power.

(B) This incorrect answer results when the initial pressure (atmospheric, 14.7 psi) is multiplied by the compression ratio instead of divided by it.

(C) This incorrect answer results when an incorrect ratio of specific heats of 1.3 is used.

SOLUTION 37

The Mach number can be determined using the isentropic flow table if the isentropic flow factor p/p_0 is known. The stagnation pressure, or total pressure, p_0, is given as 500 kPa.

If $p_{\text{ambient}}/p_{\text{reservoir}}$ is less than a critical value, then the exit static pressure will be the condition that determines the speed of the flow. The exit static pressure is nearly equal to the atmospheric pressure.

$$p = p_e \approx p_{\text{ambient}} = 101.3 \; \text{kPa}$$

The isentropic flow factor is

$$\frac{p}{p_0} = \frac{101.3 \; \text{kPa}}{500 \; \text{kPa}} = 0.2026$$

From the isentropic flow table, an isentropic flow factor of 0.2026 corresponds to a Mach number of 1.7. [One-Dimensional Isentropic Compressible-Flow Functions $k = 1.4$]

The answer is (B).

Why Other Options Are Wrong

(A) This incorrect answer results when the exit pressure is equal to the reservoir pressure.

(C) This incorrect answer results when the value 0.2026 is found in the ρ/ρ_0 column of the isentropic flow table.

(D) This incorrect answer results when the value 0.2026 is found in the T/T_0 column of the isentropic flow table.

SOLUTION 38

Since the mass flow rate remains constant, the continuity equation can be used to find the velocity at the downstream point.

Continuity Equation

$$\dot{m} = \rho A \mathrm{v}$$
$$\dot{m}_u = \dot{m}_d$$
$$\rho_u A_u \mathrm{v}_u = \rho_d A_d \mathrm{v}_d$$

The pipe area is constant, so the equation simplifies to

$$\rho_u \mathrm{v}_u = \rho_d \mathrm{v}_d$$

The densities are found using the ideal gas law.

$$p_u V_u = m_u R T_u$$
$$\rho_u = \frac{p_u}{R T_u}$$

The specific gas constant of CO_2 is 35.1 ft-lbf/lbm-°R. [Fundamental Constants]

$$\rho_u = \frac{\left(40 \ \dfrac{\mathrm{lbf}}{\mathrm{in}^2}\right)\left(12 \ \dfrac{\mathrm{in}}{\mathrm{ft}}\right)^2}{\left(35.1 \ \dfrac{\mathrm{ft\text{-}lbf}}{\mathrm{lbm\text{-}°R}}\right)(80°F + 460°)} = 0.304 \ \mathrm{lbm/ft}^3$$

$$\rho_d = \frac{\left(30 \ \dfrac{\mathrm{lbf}}{\mathrm{in}^2}\right)\left(12 \ \dfrac{\mathrm{in}}{\mathrm{ft}}\right)^2}{\left(35.1 \ \dfrac{\mathrm{ft\text{-}lbf}}{\mathrm{lbm\text{-}°R}}\right)(100°F + 460°)} = 0.22 \ \mathrm{lbm/ft}^3$$

Applying the continuity equation,

$$\left(0.304 \ \frac{\mathrm{lbm}}{\mathrm{ft}^3}\right)\left(20 \ \frac{\mathrm{ft}}{\mathrm{sec}}\right) = \left(0.22 \ \frac{\mathrm{lbm}}{\mathrm{ft}^3}\right)\mathrm{v}_d$$
$$\mathrm{v}_d = 27.6 \ \mathrm{ft/sec}$$

The flow at the downstream point is

Continuity Equation

$$Q_d = A_d \mathrm{v}_d$$
$$= \left(\frac{\pi}{4}(6 \ \mathrm{in})^2\left(\frac{1 \ \mathrm{ft}}{12 \ \mathrm{in}}\right)^2\right)\left(27.6 \ \frac{\mathrm{ft}}{\mathrm{sec}}\right)$$
$$= 5.4 \ \mathrm{ft}^3/\mathrm{sec}$$

The answer is (B).

Why Other Options Are Wrong

(A) This incorrect answer results when 15 ft/sec is used as the velocity.

(C) This incorrect answer results when the temperatures are not converted to Rankine.

(D) This incorrect answer is the velocity of the CO_2 at the downstream point.

SOLUTION 39

To create dynamic similarity between the prototype and the model, the Reynolds numbers must be equal. Applying the equation for the Reynolds number, this means that the relationship shown must be true.

Reynolds Number

$$\mathrm{Re} = \frac{\mathrm{v} D}{\nu}$$
$$\mathrm{Re}_p = \mathrm{Re}_m$$
$$\frac{\mathrm{v}_p D_p}{\nu_p} = \frac{\mathrm{v}_m D_m}{\nu_m}$$

The velocities and diameters correspond to the inlet velocity and diameter, respectively. Assuming the water has the same properties in both the prototype and the model, $\upsilon_m = \upsilon_p$, the relationship can be simplified.

$$\mathrm{v}_p D_p = \mathrm{v}_m D_m$$

By the continuity equation, the velocity is equal to

Continuity Equation

$$Q = A \mathrm{v}$$
$$\mathrm{v} = \frac{Q}{A}$$

Inserting this into the relationship and rearranging, the required diameter is

$$\frac{Q_p D_p}{A_p} = \frac{Q_m D_m}{A_m}$$

$$\frac{Q_p D_p}{\left(\frac{\pi}{4}\right) D_p^2} = \frac{Q_m D_m}{\left(\frac{\pi}{4}\right) D_m^2}$$

$$\frac{Q_p}{D_p} = \frac{Q_m}{D_m}$$

$$D_m = \frac{Q_m}{Q_p} D_p$$

$$= \left(\frac{5 \ \frac{\text{ft}^3}{\text{sec}}}{40 \ \frac{\text{ft}^3}{\text{sec}}}\right)(2 \ \text{ft})\left(\frac{12 \ \text{in}}{1 \ \text{ft}}\right)$$

$$= 3.0 \ \text{in}$$

The answer is (B).

Why Other Options Are Wrong

(A) This incorrect answer results from not converting feet to inches.

(C) This incorrect answer results when areas, not diameters, are used.

(D) This incorrect answer results when the ratio Q_p/Q_m is used and the solution is not converted to inches.

SOLUTION 40

For schedule-40 pipe, use a table to find the areas of the different pipes. [Schedule 40 Steel Pipe]

$$A_A = A_B = \frac{0.304 \ \text{in}^2}{\left(12 \ \frac{\text{in}}{\text{ft}}\right)^2} = 0.00211 \ \text{ft}^2$$

$$A_C = \frac{0.533 \ \text{in}^2}{\left(12 \ \frac{\text{in}}{\text{ft}}\right)^2} = 0.00370 \ \text{ft}^2$$

The water velocity at location C is found using the continuity equation.

Continuity Equation

$$Q = Av$$

$$Q_C = Q_A + Q_B$$

$$A_C v_C = A_A v_A + A_B v_B$$

$$v_C = \frac{A_A v_A + A_B v_B}{A_C}$$

$$= \frac{\begin{array}{c}(0.00211 \ \text{ft}^2)\left(12 \ \frac{\text{ft}}{\text{sec}}\right) \\ + (0.00211 \ \text{ft}^2)\left(25 \ \frac{\text{ft}}{\text{sec}}\right)\end{array}}{0.00370 \ \text{ft}^2}$$

$$= 21.10 \ \text{ft/sec}$$

Draw a free-body diagram of the pipe at location C.

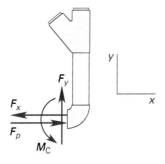

The steady-flow equation in the x-direction is

Impulse-Momentum Principle

$$\sum F_x = \sum Q_2 \rho_2 v_{2x} - \sum Q_1 \rho_1 v_{1x}$$

$$= \sum (Av)_2 \rho_2 v_{2x} - \sum (Av)_1 \rho_1 v_{1x}$$

The densities at locations A, B, and C are the same. This equation becomes

$$\sum F_x = \sum \frac{(Av)_2 \rho v_{2x}}{g_c} - \sum \frac{(Av)_1 \rho v_{1x}}{g_c}$$

$$= \left(\frac{\rho A v}{g_c}\right)_A v_{A,x} + \left(\frac{\rho A v}{g_c}\right)_B v_{B,x} - \left(\frac{\rho A v}{g_c}\right)_C v_{C,x}$$

The force due to the static pressure at location C is

$$F_p = p A_C = \left(40 \ \frac{\text{lbf}}{\text{in}^2}\right)(0.533 \ \text{in}^2)$$

$$= 21.32 \ \text{lbf}$$

Summing forces in the x-direction at location C, the equation becomes

$$F_p - R_x = \left(\frac{\rho v A}{g_c}\right)_A v_{A,x} + \left(\frac{\rho v A}{g_c}\right)_B v_{B,x} - \left(\frac{\rho v A}{g_c}\right)_C v_{C,x}$$

The horizontal force needed is

$$F_R = F_p - \left(\frac{\rho}{g_c}\right)$$

$$\times \left(\begin{array}{c} (vA)_A v_{A,x} + (vA)_B v_{B,x} \\ -(vA)_C v_{C,x} \end{array} \right)$$

$$= 21.32 \text{ lbf} - \left(\frac{62.4 \dfrac{\text{lbm}}{\text{ft}^3}}{32.17 \dfrac{\text{lbm-ft}}{\text{lbf-sec}^2}} \right)$$

$$\times \left(\begin{array}{c} \left(12 \dfrac{\text{ft}}{\text{sec}}\right)(0.00211 \text{ ft}^2) \\ \times \left(\left(-12 \dfrac{\text{ft}}{\text{sec}}\right)\sin 30°\right) \\ +\left(25 \dfrac{\text{ft}}{\text{sec}}\right)(0.00211 \text{ ft}^2) \\ \times \left(0 \dfrac{\text{ft}}{\text{sec}}\right) \\ -\left(21.10 \dfrac{\text{ft}}{\text{sec}}\right)(0.00370 \text{ ft}^2) \\ \times \left(21.10 \dfrac{\text{ft}}{\text{sec}}\right) \end{array} \right)$$

$$= 24.81 \text{ lbf} \quad (25 \text{ lbf})$$

The answer is (D).

Why Other Options Are Wrong

(A) This incorrect answer results when the static pressure is not included in the force calculation.

(B) This incorrect answer results when an incorrect pipe area at A and B is used to compute the static pressure force.

(C) This incorrect answer is the force due to the static pressure alone.

SOLUTION 41

The pressure change when the valve is closed is found using the equation for water hammer.

Water Hammer

$$\Delta p_h = \frac{\rho C_s v}{g_c}$$

At 60°F, the density of water, ρ, is 62.4 lbm/ft³. [Properties of Water at Standard Conditions]

Calculate the speed of sound, c, in water.

Mach Number

$$c = \sqrt{\frac{Bg_c}{\rho}}$$

$$= \sqrt{\frac{\left(300{,}000 \dfrac{\text{lbf}}{\text{in}^2}\right)\left(12 \dfrac{\text{in}}{\text{ft}}\right)^2\left(32.17 \dfrac{\text{lbm-ft}}{\text{lbf-sec}^2}\right)}{62.4 \dfrac{\text{lbm}}{\text{ft}^3}}}$$

$$= 4721 \text{ ft/sec}$$

The change in pressure is

$$\Delta p = \frac{\rho c \Delta v}{g_c}$$

$$= \frac{\left(62.4 \dfrac{\text{lbm}}{\text{ft}^3}\right)\left(4721 \dfrac{\text{ft}}{\text{sec}}\right)\left(3 \dfrac{\text{ft}}{\text{sec}}\right)}{\left(32.17 \dfrac{\text{lbm-ft}}{\text{lbf-sec}^2}\right)\left(12 \dfrac{\text{in}}{\text{ft}}\right)^2}$$

$$= 190.6 \text{ lbf/in}^2$$

The maximum pressure is

$$p_{\max} = p_0 + \Delta p = 60 \frac{\text{lbf}}{\text{in}^2} + 190.6 \frac{\text{lbf}}{\text{in}^2}$$

$$= 250.6 \text{ lbf/in}^2 \quad (250 \text{ psi})$$

The answer is (D).

Why Other Options Are Wrong

(A) This incorrect answer results when the flow rate is used instead of the velocity.

(B) This incorrect answer results when the speed of sound in meters per second is used instead of feet per second.

(C) This incorrect answer is the change in pressure alone and with no safety factor included.

SOLUTION 42

From the graph, the pump efficiency at 400 gal/min is approximately 60%, and the amount of head gained is approximately 590 ft. The amount of pump head needed at the shaft is

$$h_{\text{shaft}} = \frac{h_{\text{gained}}}{\eta} = \frac{590 \text{ ft}}{0.6} = 983 \text{ ft} \quad (980 \text{ ft})$$

The answer is (B).

Why Other Options Are Wrong

(A) This incorrect answer results when the maximum pump efficiency (approximately 70%) is used instead of the pump efficiency at the operating point.

(C) This incorrect answer results when the head value used is taken at the point where the vertical 400 gal/min line crosses the efficiency curve (approximately 650 ft).

(D) This incorrect answer results when an efficiency of 40% (that is, 100% − 60%) is used instead of 60%.

SOLUTION 43

The turbine expands the steam isentropically, so

$$s_2 = s_1 = 1.612 \text{ Btu/lbm-°R}$$

The entropies of saturated liquid at 2 psi are found in a table of properties of saturated water and steam. [Properties of Saturated Water and Steam (Pressure) - I-P Units]

$$s_f = 0.1750 \text{ Btu/lbm-°R}$$
$$s_{fg} = 1.7445 \text{ Btu/lbm-°R}$$

The steam quality at state 2 is

Properties for Two-Phase (Vapor-Liquid) Systems

$$s_2 = s_f + x_2 s_{fg}$$

$$x_2 = \frac{s_2 - s_f}{s_{fg}}$$

$$= \frac{1.612 \dfrac{\text{Btu}}{\text{lbm-°R}} - 0.1750 \dfrac{\text{Btu}}{\text{lbm-°R}}}{1.7445 \dfrac{\text{Btu}}{\text{lbm-°R}}}$$

$$= 0.8237$$

The enthalpy at state 3, h_3, is given as 94 Btu/lbm. The enthalpies of saturated liquid at 2 psi can be found in a table of properties of saturated water and steam. [Properties of Saturated Water and Steam (Pressure) - I-P Units]

$$h_f = 94.01 \text{ Btu/lbm}$$
$$h_{fg} = 1021.74 \text{ Btu/lbm}$$

The enthalpy at state 2 is

Properties for Two-Phase (Vapor-Liquid) Systems

$$h_2 = h_f + x_2 h_{fg}$$

$$= 94.01 \frac{\text{Btu}}{\text{lbm}} + (0.8237)\left(1021.74 \frac{\text{Btu}}{\text{lbm}}\right)$$

$$= 935.62 \text{ Btu/lbm}$$

The heat flow per unit mass from the condenser is

$$Q_L = h_2 - h_3$$
$$= 935.62 \frac{\text{Btu}}{\text{lbm}} - 94.01 \frac{\text{Btu}}{\text{lbm}}$$
$$= 841.61 \text{ Btu/lbm} \quad (840 \text{ Btu/lbm})$$

The answer is (B).

Why Other Options Are Wrong

(A) This incorrect answer is the enthalpy at state 3.

(C) This incorrect answer results when a steam quality of 100% is used.

(D) This incorrect answer results when the enthalpy at state 1 is used in the final calculation instead of the enthalpy at state 2.

SOLUTION 44

Statements I and III are true. Gas turbine power cycles typically operate at higher temperatures than steam power cycles, so statement II is false. The practical upper limit for the efficiency of a combined power cycle is considered to be approximately 70%, so statement IV is false.

The answer is (A).

Why Other Options Are Wrong

(B) This answer is wrong because item II is false. Gas turbine power cycles operate at higher temperatures than steam power cycles.

(C) This answer is wrong because item II is false. Gas turbine power cycles operate at higher temperatures than steam power cycles.

(D) This answer is wrong because the practical upper limit for the efficiency of a combined power cycle is approximately 70%.

2 Hydraulic and Fluid Applications

PROBLEM 1

A 150 hp reduced crude pump in a refinery pumps 1700 gal/min. The brake horsepower needed to pump 2500 gal/min is most nearly

(A) 150 hp

(B) 220 hp

(C) 320 hp

(D) 480 hp

Hint: The relationship between the brake horsepower of a pump and its flow rate is nonlinear.

PROBLEM 2

A lake supplies water to drive an impulse turbine that uses blades with a 165° exit angle. The combined friction and minor head loss in the pipe is 500 ft. The pipe section is at an elevation of 2000 ft.

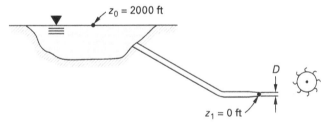

The nozzle diameter that will provide a maximum power of 2.5 MW is most nearly

(A) 0.29 in

(B) 3.4 in

(C) 5.8 in

(D) 11 in

Hint: Maximum power occurs when $v = v_1/2$.

PROBLEM 3

A centrifugal pump at an oil pipeline pumping station pumps 70,000 barrels per day of crude oil through a 12 in schedule-40 steel pipe to a booster pump inside a refinery 10 mi away. The crude oil has a specific gravity of 0.86 and a kinematic viscosity of 250×10^{-6} ft^2/sec. The pump develops 956 ft of head. The motor torque supplied to the pump shaft running at 1750 rpm is 1500 lbf-ft. The pump's efficiency is most nearly

(A) 14%

(B) 67%

(C) 85%

(D) 98%

Hint: Calculate the brake horsepower of the pump.

PROBLEM 4

The feedwater piping system of a boiler is designed for an operating pressure of 2200 psi and a temperature of 280°F at the outlet. The friction pressure drop in the piping system under the design conditions is 33 psi. The design flow rate is 1.4×10^6 lbm/hr. The velocity pressure differentials and the static pressure drops are negligible; the lowest mass flow rate is 740,000 lbm/hr. The lowest pressure drop in the feedwater piping is most nearly

(A) 9.2 lbf/in^2

(B) 17 lbf/in^2

(C) 33 lbf/in^2

(D) 120 lbf/in^2

Hint: Assume the friction pressure drop is proportional to the square of the velocity.

PROBLEM 5

A pipeline carries crude oil that has a specific gravity of 0.86, and a kinematic viscosity of 250×10^{-6} ft^2/sec. A 250 hp pump at the pipeline pumping station generates 450 ft of head. The increase in pressure at the pump discharge is most nearly

(A) 170 psi

(B) 200 psi

(C) 2000 psi

(D) 24,000 psi

Hint: Equate feet of head to static pressure.

PROBLEM 6

A Carnot engine operates with a minimum temperature of 80°F, a maximum temperature of 1200°F, and a maximum pressure of 1300 psia. The working fluid is 0.2 lbm of air, and 5 Btu of heat is added to the air. The ratio of specific heats for the air is 1.4. Assume the air behaves as an ideal gas. The equation for the flow of the heat into the engine is

$$q_{\text{in}} = p_2 V_2 \ln \frac{V_3}{V_2}$$

The maximum cylinder volume is most nearly

(A) 0.10 ft^3

(B) 1.5 ft^3

(C) 2.0 ft^3

(D) 2.7 ft^3

Hint: Begin at the state where the pressure is a maximum.

PROBLEM 7

An air compressor operates at steady state. Air enters through a 1 ft^2 opening at a velocity of 20 ft/sec, a temperature of 77°F, a pressure of 14.7 psia, and an enthalpy of 128.34 Btu/lbm. The air exits at a velocity of 6 ft/sec, a temperature of 360°F, a pressure of 100 psia, and an enthalpy of 196.69 Btu/lbm. The power input to the compressor is 150 hp. The sums of the mass input and exit flow rates are equal.

With ideal air properties, the heat input to the compressor is most nearly

(A) −48,000 Btu/hr

(B) −19,000 Btu/hr

(C) −18,000 Btu/hr

(D) 0 Btu/hr

Hint: Neglect potential energy changes.

PROBLEM 8

A Carnot cycle undergoes isothermal compression at 100°F and isothermal expansion at 300°F. The thermal efficiency of this cycle is most nearly

(A) 26%

(B) 35%

(C) 67%

(D) 75%

Hint: Internal combustion engines are evaluated using the Otto cycle.

PROBLEM 9

Which of the following statements about the thermal efficiency of reversible heat engines are true?

I. The higher the temperature at which heat is added, the higher the efficiency.

II. The maximum thermal efficiency is achieved using a Carnot cycle.

III. The lower the temperature at which heat is rejected, the higher the efficiency.

IV. Different reversible engines operating between the same temperature limits have different thermal efficiencies.

(A) I and II

(B) I and IV

(C) II and III

(D) I, II, and III

Hint: Heat engines are thermodynamic cycles.

PROBLEM 10

A pump rated for 350 gal/min and 200 ft of head is needed for use as a refinery boiler feedwater pump. The pump will run continuously year-round. The manufacturer's performance curves for a particular pump are shown in *Illustration for Problem 10*.

The pump is oversized by 20%, and the increased electrical cost is $0.07/kW-hr. The yearly electrical cost increase for the oversized pump is most nearly

(A) $0

(B) $915

(C) $2750

(D) $5490

Hint: Be sure to increase both the flow rate and head by 20%. Then treat the oversize requirement as a new rating and design accordingly.

PROBLEM 11

A refinery boiler feedwater pump operating at 3550 rpm has a 9.5 in impeller. The manufacturer's performance curves are shown in *Illustration for Problem 11*.

The pump needs to achieve a rating of 550 gal/min and 350 ft of head. The speed at which the pump should be run is most nearly

(A) 3070 rpm

(B) 4150 rpm

(C) 4740 rpm

(D) 5140 rpm

Hint: Use trial and error.

PROBLEM 12

A 2.5 in diameter solid steel shaft turns a cooling tower fan at 200 rpm. The maximum horsepower that can be transmitted is most nearly

(A) 80 hp

(B) 150 hp

(C) 190 hp

(D) 290 hp

Hint: Use the maximum shear stress failure criterion.

PROBLEM 13

A crude overhead product pump operating at 3550 rpm has the characteristics shown.

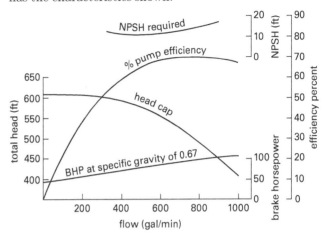

The pump is connected to a system requiring 200 ft of head when handling 500 gal/min of crude oil. The system curve can be approximated as follows.

$$h_1 = h_2 \left(\frac{Q_1}{Q_2} \right)^2$$

The flow rate in the system when the pump is running at 3550 rpm is most nearly

(A) 220 gal/min

(B) 300 gal/min

(C) 800 gal/min

(D) 1000 gal/min

Hint: Draw the system curve.

Illustration for Problem 10

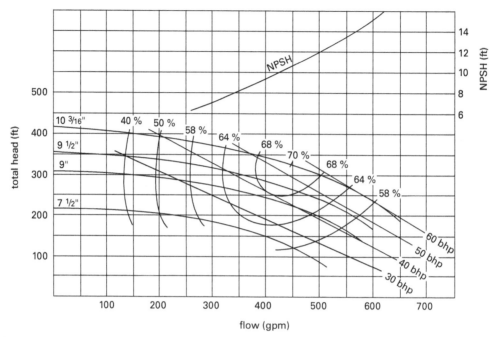

Reprinted/adapted with permission from Flowserve.

Illustration for Problem 11

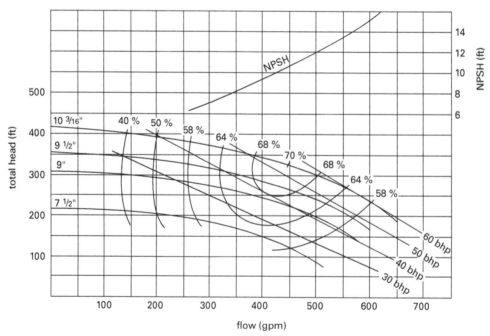

Reprinted/adapted with permission from Flowserve.

PROBLEM 14

The turbine shown has an output of 10 MW and an efficiency of 90%. Saturated steam at 1000 psia flows from the boiler through the turbine and exhausts to a condenser at 2 psia. The condensate is subcooled 6°F in the condenser. The flow rate of the condensate is 100,000 lbm/hr.

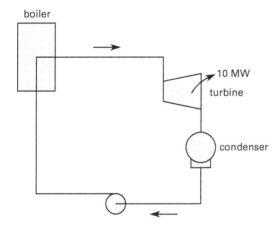

The thermal efficiency of the cycle is most nearly

(A) 9.1%

(B) 28%

(C) 31%

(D) 34%

Hint: Sufficient accuracy can be obtained by neglecting pump work.

PROBLEM 15

A steam turbine operates at an overall efficiency of 75% with a back pressure of 300 psia. The system uses superheated steam at 1300 psia and 950°F. The amount of steam needed to generate a power of 9 MW is most nearly

(A) 66.5×10^3 lbm/hr

(B) 113×10^3 lbm/hr

(C) 182×10^3 lbm/hr

(D) 220×10^3 lbm/hr

Hint: The expansion inside a turbine can be modeled as an isentropic process.

PROBLEM 16

Water at 180°F is pumped out of a vented tank at 140 gal/min. The pump manufacturer requires 5 ft of net positive suction head (NPSH) for the pump at this flow. Surroundings are at 1.0 atm.

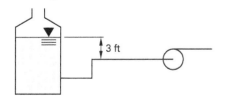

The head loss between the tank and the pump suction is given by the equation

$$h_f = 1.50 \times 10^{-5} L Q^{1.852}$$

h_f is the head loss due to friction in feet, L is the pipeline length in feet, and Q is the flow rate in gal/min. The maximum possible pipeline length without cavitation is most nearly

(A) 85 ft

(B) 100 ft

(C) 140 ft

(D) 230 ft

Hint: Compare required NPSH to available NPSH.

PROBLEM 17

An impeller with a 2.0 ft diameter operates at its best efficiency point. At this efficiency point, its speed is 440 rpm, and it produces a flow rate of 200 gal/min, a head of 180 ft, and 17.0 hp of power. The impeller is clipped to 1.5 ft and still operates at 440 rpm. The new flow rate, head, and brake horsepower will be most nearly

(A) 150 gal/min, 140 ft, and 13 hp

(B) 200 gal/min, 140 ft, and 9.6 hp

(C) 84 gal/min, 76 ft, and 7.2 hp

(D) 150 gal/min, 100 ft, and 7.2 hp

Hint: The differences between clipped and non-clipped impellers are small.

PROBLEM 18

An industrial-grade drum fan has three blades and rotates at a speed of 490 rpm. The blades have a diameter of 42 in with an effective pitch of 2.75 ft per revolution. The airflow of the fan is most nearly

(A) 220 ft³/min

(B) 3700 ft³/min

(C) 13,000 ft³/min

(D) 17,000 ft³/min

Hint: Think of a screw pitch.

PROBLEM 19

The flow characteristics of a typical centrifugal compressor are shown.

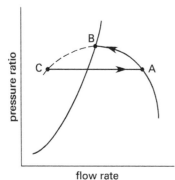

The cycle ABCA is the phenomenon known as

(A) surging

(B) stalling

(C) starving

(D) throttling

Hint: Look carefully at the flow rate during this cycle.

PROBLEM 20

Kerosene (SG = 0.85) flows through a Venturi meter at 800 gal/min as shown in the figure. The cross-sectional area of the flow entering the meter is 0.085 ft², and the cross-sectional area of the flow exiting the meter is 0.030 ft².

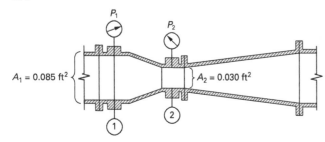

The pressure difference needed to measure this flow rate is most nearly

(A) 0.18 psi

(B) 10.7 psi

(C) 13.0 psi

(D) 17.7 psi

Hint: Assume C_v = 1.0.

PROBLEM 21

A 44 in long, 1.25 in diameter compressor shaft is positioned in v-blocks. A dial indicator is placed on the shaft to determine the shaft runout. The readings of the indicator are given in the following table.

angle (deg)	reading (0.1 mil)	angle (deg)	reading (0.1 mil)
0	0.0	190	1.0
10	0.0	200	0.5
20	0.0	210	0.5
30	0.5	220	0.5
40	0.5	230	0.0
50	0.5	240	0.0
60	0.5	250	0.0
70	1.0	260	−0.5
80	1.0	270	−0.5
90	1.0	280	−0.5
100	1.0	290	−0.5
110	1.0	300	−1.0
120	1.5	310	−1.0
130	1.5	320	−0.5
140	1.5	330	−0.5
150	1.5	340	−0.5
160	1.5	350	0.0
170	1.0	360	0.0
180	1.0		

The mechanical runout of the shaft is most nearly

(A) 0.036 mil

(B) 0.10 mil

(C) 0.15 mil

(D) 0.25 mil

Hint: The mechanical runout is a difference.

PROBLEM 22

A pitot-static tube on a formula race car measures dynamic pressure while the car races in still air at standard temperature and pressure. The speed of the car at a dynamic pressure of 7.0 in of water is most nearly

(A) 20 mi/hr

(B) 45 mi/hr

(C) 120 mi/hr

(D) 144 mi/hr

Hint: Look carefully at the units of velocity.

PROBLEM 23

A three-phase motor operating at 93% efficiency and pulling 100 A drives a feedwater pump for a high-pressure boiler in a refinery. The flow rate of water through the pump is 500 gal/min, and there is a pressure differential of 850 psi.

The equation for the power output (in kilowatts) of a motor is

$$P_{kW} = \frac{\eta_m \sqrt{3}\, IV(\text{pf})}{1000}$$

The motor voltage is 2300 V, and the power factor is 0.87. The pump efficiency is most nearly

(A) 57%

(B) 77%

(C) 86%

(D) 99%

Hint: When efficiencies are given, begin looking at power.

PROBLEM 24

A run-around heat recovery system uses a coolant loop and two crossflow heat exchangers to extract energy from warm air exhausted from a building and transfer the energy to cold air brought into the building.

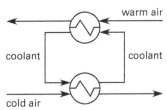

Before passing through the system, the warm air is at 75°F, and the cold air is at 10°F. Air flows through the system in both directions at 600 lbm/min, and coolant flows at 250 lbm/min. Heat is removed from the outgoing air at 200,000 Btu/hr. The overall heat transfer coefficient for each heat exchanger is 50 Btu/hr-ft²-°F, and the fluids are not mixed. The specific heats are 0.24 Btu/lbm-°F for the air and 1.0 Btu/lbm-°F for the coolant. Heat loss from the coolant line and energy added by the coolant pump can be neglected. The coolant temperatures in the left and right halves of the loop, respectively, are most nearly

(A) 30°F and 54°F

(B) 49°F and 36°F

(C) 54°F and 30°F

(D) 75°F and 10°F

Hint: Due to the symmetry of this arrangement, the average of the coolant temperatures on the two sides of the loop are equal to the average of the temperatures of the incoming warm and cold air.

PROBLEM 25

A carbon dioxide (CO_2) sequestration pipeline uses a schedule-40 pipe with a 6 in inner diameter. At point A, CO_2 flows through the pipe at a speed of 20 ft/sec with a pressure of 40 psig and a temperature of 70°F. Downstream at point B, the pressure is 30 psig, and the temperature is 90°F. The flow rate at point B is most nearly

(A) 5 ft³/min

(B) 236 ft³/min

(C) 300 ft³/min

(D) 326 ft³/min

Hint: Remember that CO_2 is compressible.

PROBLEM 26

Water enters a boiler at 90°F. Gas enters the boiler at 1300°F and leaves at 215°F, increasing the temperature of the water in the boiler to 212°F. The log mean temperature difference is most nearly

(A) 80.0°F

(B) 184°F

(C) 446°F

(D) 758°F

Hint: This is similar to a heat exchanger problem.

PROBLEM 27

An air-water-vapor mixture exists at 30 psia, 150°F, and 75% relative humidity. The actual partial vapor pressure is 2.79 psia. The humidity ratio is most nearly

(A) 0.064 $\text{lbm}_{\text{vapor}}/\text{lbm}_{\text{air}}$

(B) 0.077 $\text{lbm}_{\text{vapor}}/\text{lbm}_{\text{air}}$

(C) 0.165 $\text{lbm}_{\text{vapor}}/\text{lbm}_{\text{air}}$

(D) 0.243 $\text{lbm}_{\text{vapor}}/\text{lbm}_{\text{air}}$

Hint: The specific gas constant, R, for water vapor is 85.78 ft-lbf/lbm-°R.

PROBLEM 28

A noncooled main air blower at a refinery compresses 50,000 cfm of atmospheric air to 53 psia. The compressor efficiency is 79%. The motor shaft horsepower needed to operate the air blower is most nearly

(A)   110 hp

(B)   1800 hp

(C)   4000 hp

(D)   6300 hp

Hint: Think adiabatic power.

PROBLEM 29

Water flows from a small upper tank to a small lower tank at the rate of 60 cfm in a smooth pipe with a 6 in inside diameter.

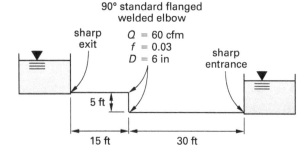

The friction factor is 0.03. The total head loss through the pipe is most nearly

(A)   0.4 ft

(B)   0.9 ft

(C)   1.3 ft

(D)   2.1 ft

Hint: Be sure to include all the head loss terms.

PROBLEM 30

A centrifugal pump is placed above a large reservoir and will be used to pump 10 gal/min of water. To prevent cavitation at this flow rate, the pump manufacturer specifies that the pump inlet pressure should be 20 psi. The water temperature is 68°F, and atmospheric pressure is 14.7 psi. The net positive suction head required is most nearly

(A)   3.8 ft

(B)   20 ft

(C)   46 ft

(D)   80 ft

Hint: Remember to do all the conversions needed.

PROBLEM 31

An AC three-phase generator supplies 20 A at 240 V. The lagging power factor is 0.8, and the generator efficiency is 90%. The torque needed to drive the generator at 1800 rpm is most nearly

(A)   17 lbf-ft

(B)   29 lbf-ft

(C)   37 lbf-ft

(D)   50 lbf-ft

Hint: Be sure to realize that this is a three-phase generator.

PROBLEM 32

Desert air at 115°F and 10% relative humidity is to be cooled using an evaporative cooler. The minimum temperature to which the air can be cooled is most nearly

(A)   35°F

(B)   45°F

(C)   71°F

(D)   85°F

Hint: A chart from ASHRAE will help.

PROBLEM 33

A concrete pump is pumping concrete at a flow rate of 1.20 ft³/sec with a pressure of 420 lbf/in². The specific gravity of the concrete is 2.0. The pump has an efficiency of 60%. The hydraulic horsepower is most nearly

(A)   66 hp

(B)   110 hp

(C)   130 hp

(D)   230 hp

Hint: Be sure to know the difference between hydraulic (water) horsepower and brake horsepower.

PROBLEM 34

An airplane flies straight and level at 10,000 ft and 140 mph. The airplane has a wing surface area of 174 ft², and the mass of the airplane is 2450 lbm. The coefficient of lift varies linearly from 0.25 at 0° to 0.75 at 5°. The angle of attack of the wings is most nearly

(A) 0°

(B) 0.42°

(C) 1.7°

(D) 4.7°

Hint: Interpolation will be necessary.

PROBLEM 35

A 2.5 lbm mass, 6 in diameter sphere is submerged in oil with a density of 56 lbm/ft^3. The sphere rises with a constant velocity. If the drag coefficient is 0.5, the velocity is most nearly

(A) 0.5 ft/sec

(B) 3.7 ft/sec

(C) 6.6 ft/sec

(D) 10 ft/sec

Hint: A free-body diagram will be useful.

PROBLEM 36

A lock on the Volga River in Russia has surface dimensions of 950 ft by 100 ft and a depth of 36 ft. The area of the opening into the lock from the river is 140 ft^2. The lock and canal can be modeled as shown.

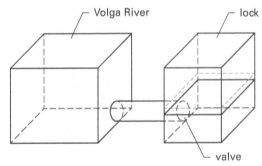

When the lock is half-filled, the velocity at which the surface of the water in the lock will rise is most nearly

(A) 0.6 ft/min

(B) 3.0 ft/min

(C) 30 ft/min

(D) 100 ft/min

Hint: The surface area of the lock is much larger than the surface area of the opening into the lock from the valve.

PROBLEM 37

Three oil derricks each produce the same number of barrels of oil per day. The output from each derrick is sent to a main transportation line by way of a feeder line constructed of 16 in schedule-40 pipe, which has an internal diameter of 15 in.

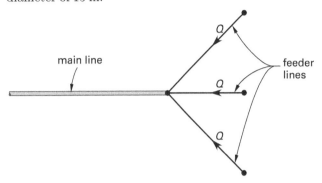

The main line is constructed of 30 in schedule-40 pipe, which has an internal diameter of 29 in. Elevation effects are negligible, and the friction factors in all the pipes are equal. The ratio of the pressure drop per unit length in the main transportation line to that in the feeder lines is most nearly

(A) 0.04

(B) 0.33

(C) 1.4

(D) 3.0

Hint: Don't forget about the flow rate equation $Q = Av$.

SOLUTION 1

This problem is solved using the pump affinity laws.

Pump Affinity Laws

$$Q_2 = Q_1 \left(\frac{N_2}{N_1} \right)$$

$$\text{bhp}_2 = \text{bhp}_1 \left(\frac{N_2}{N_1} \right)^3$$

Substitute the flow rate ratio for the speed ratio in the brake horsepower equation, substitute the given values, and solve for the new brake horsepower.

$$\text{bhp}_2 = \text{bhp}_1 \left(\frac{Q_2}{Q_1} \right)^3$$

$$= (150 \text{ hp}) \left(\frac{2500 \text{ gal/min}}{1700 \text{ gal/min}} \right)^3$$

$$= 477 \text{ hp} \quad (480 \text{ hp})$$

The answer is (D).

Why Other Options Are Wrong

(A) This incorrect answer results when the brake horsepower of the pump is not understood to be a function of the pump's flow rate.

(B) This incorrect answer results when the flow rate ratio is not cubed in the calculation of the brake horsepower.

(C) This incorrect answer results when the flow rate ratio is squared instead of cubed in the calculation of the brake horsepower.

SOLUTION 2

The power developed from an impulse turbine is given by

Impulse Turbine

$$\dot{W} = Q \left(\frac{\rho}{g} \right) (v_1 - v)(1 - \cos\alpha) v$$

Convert $\dot{W}$ into ft-lbf/sec.

$$\dot{W} = (2.5 \text{ MW}) \left(\frac{1 \times 10^6 \text{ W}}{1 \text{ MW}} \right) \left(\frac{1 \text{ hp}}{745.7 \text{ W}} \right) \left(\frac{550 \frac{\text{ft-lbf}}{\text{sec}}}{1 \text{ hp}} \right)$$

$$= 1.84 \times 10^6 \text{ ft-lbf/sec}$$

The velocity, v_1, exiting the nozzle can be found using the Bernoulli equation.

Bernoulli Equation

$$\frac{p_0}{\gamma} + z_0 + \frac{v_0^2}{2g} = \frac{p_1}{\gamma} + z_1 + \frac{v_1^2}{2g} + h_f$$

The pressure at points 0 and 1 is atmospheric, so $p_0 = p_1 = 0$. All heights are referenced from point 1, so $z_1 = 0$. The surface area of the lake is large, so $v_0 = 0$. The Bernoulli equation can thus be simplified. The velocity exiting the nozzle is

$$v_1 = \sqrt{2g(z_0 - h_f)}$$

$$= \sqrt{(2) \left(32.17 \frac{\text{ft}}{\text{sec}^2} \right) (2000 \text{ ft} - 500 \text{ ft})}$$

$$= 311 \text{ ft/sec}$$

Since the maximum power occurs when $v = v_1/2$, substitute $v_1/2$ for v when solving.

$$\dot{W} = Q \left(\frac{\rho}{g_c} \right) \left(v_1 - \frac{v_1}{2} \right) (1 - \cos\alpha) \left(\frac{v_1}{2} \right)$$

$$1.84 \times 10^6 \frac{\text{ft-lbf}}{\text{sec}} = Q \left(\frac{62.4 \frac{\text{lbm}}{\text{ft}^3}}{32.17 \frac{\text{lbm-ft}}{\text{lbf-sec}^2}} \right) \left(v_1 - \frac{v_1}{2} \right)$$

$$\times (1 - \cos 165°) \left(\frac{v_1}{2} \right)$$

$$1.84 \times 10^6 \frac{\text{ft-lbf}}{\text{sec}} = Q \left(3.81 \frac{\text{lbf-sec}^2}{\text{ft}^4} \right) \left(\frac{v_1^2}{4} \right)$$

Using the continuity equation, Q can be expressed as

Continuity Equation

$$Q = A v_1$$

The power equation then becomes

$$1.84 \times 10^6 \frac{\text{ft-lbf}}{\text{sec}} = (A v_1) \left(3.81 \frac{\text{lbf-sec}^2}{\text{ft}^4} \right) \left(\frac{v_1^2}{4} \right)$$

$$= \left(\frac{\pi}{4} D^2 \right) \left(3.81 \frac{\text{lbf-sec}^2}{\text{ft}^4} \right) \left(\frac{v_1^3}{4} \right)$$

$$= (D^2) \left(0.748 \frac{\text{lbf-sec}^2}{\text{ft}^4} \right) (v_1^3)$$

$$= (D^2) \left(0.748 \frac{\text{lbf-sec}^2}{\text{ft}^4} \right) \left(311 \frac{\text{ft}}{\text{sec}} \right)^3$$

Solving for D,

$$D = (0.286 \text{ ft})\left(\frac{12 \text{ in}}{1 \text{ ft}}\right) = 3.4 \text{ in}$$

The answer is (B).

Why Other Options Are Wrong

(A) This incorrect answer results when the diameter, D, is not converted to inches.

(C) This incorrect answer results when in the calculation of v_1, the factor of 2 is neglected.

(D) This incorrect answer results when 18.4×10^6 ft-lbf/sec is used instead of 1.84×10^6 ft-lbf/sec.

SOLUTION 3

The brake horsepower as a function of efficiency can be found using the pump power equation.

Pump Power Equation

$$\text{bhp} = \frac{(\text{gpm})\Delta h(\text{SG})}{3960\eta}$$

Because brake horsepower is a form of power, it can also be found from the equation for power. The brake horsepower is

$$\text{bhp} = P = T\left(2\pi \frac{\text{rad}}{\text{rev}}\right)\omega$$

$$= (1500 \text{ ft-lbf})\left(1750 \frac{\text{rev}}{\text{min}}\right)\left(2\pi \frac{\text{rad}}{\text{rev}}\right)\left(\frac{1 \text{ min}}{60 \text{ sec}}\right)$$

$$\times \left(\frac{1 \text{ hp}}{550 \frac{\text{ft-lbf}}{\text{sec}}}\right)$$

$$= 500 \text{ hp}$$

Convert the flow rate from barrels per day to gallons per minute. [Measurement Relationships]

$$Q = \left(70,000 \frac{\text{barrels}}{\text{day}}\right)\left(42 \frac{\text{gal}}{\text{barrel}}\right)$$

$$\times \left(\frac{1 \text{ day}}{1440 \text{ min}}\right)$$

$$= 2042 \text{ gal/min}$$

Solve the brake horsepower equation for the efficiency of the pump.

Pump Power Equation

$$\text{bhp} = \frac{Q\Delta h\,(\text{SG})}{3960\eta}$$

$$\eta = \frac{Q\Delta h(\text{SG})}{3960(\text{bhp})}$$

$$= \frac{\left(2042 \frac{\text{gal}}{\text{min}}\right)(956 \text{ ft})(0.86)}{(3960)(500 \text{ hp})}$$

$$= 0.848 \quad (85\%)$$

The answer is (C).

Why Other Options Are Wrong

(A) This incorrect answer results when the 70,000 barrels per day is converted to gallons per second.

(B) This incorrect answer results when 0.68 instead of 0.86 is used for the specific gravity.

(D) This incorrect answer results when the specific gravity of crude oil is not included in the water horsepower calculation. Therefore, the water horsepower is calculated for water instead of crude oil.

SOLUTION 4

The equation for the friction head at the design point can be found by simplifying the Darcy-Weisbach equation.

Head Loss Due to Flow: Darcy-Weisbach Equation

$$h_{f,d} = f\left(\frac{L}{D}\right)\left(\frac{v_d^2}{2g}\right)$$

$$= K v_d^2$$

The velocity can be expressed in terms of the mass flow rate.

Continuity Equation

$$\dot{m}_d = \rho A v_d$$

$$v_d = \frac{\dot{m}_d}{\rho A}$$

The friction head at the design point is

$$h_{f,d} = K v_d^2 = K\left(\frac{\dot{m}_d}{\rho A}\right)^2$$

Likewise, the friction head at the off-design point is

Head Loss Due to Flow: Darcy-Weisbach Equation

$$h_{f,\text{off}} = f\left(\frac{L}{D}\right)\left(\frac{v_{\text{off}}^2}{2g}\right) = K\left(\frac{\dot{m}_{\text{off}}}{\rho A}\right)^2$$

Solving these two equations for the pressure head at the off-design conditions gives

$$h_{f,\text{off}} = h_{f,d}\frac{\dot{m}_{\text{off}}^2}{\dot{m}_d^2}$$

The change in pressure in terms of head is

Bernoulli Equation

$$p_2 - p_1 = -\rho g h_f$$
$$\Delta p = -\rho g h_f$$

Combine these equations and solve for the pressure drop at the off-design conditions, which is the lowest pressure drop in the feedwater piping.

$$\Delta p_{\text{off}} = \Delta p_d\frac{\dot{m}_{\text{off}}^2}{\dot{m}_d^2}$$

$$= \left(33\ \frac{\text{lbf}}{\text{in}^2}\right)\left(\frac{\left(740{,}000\ \dfrac{\text{lbm}}{\text{hr}}\right)^2}{\left(1.4\times10^6\ \dfrac{\text{lbm}}{\text{hr}}\right)^2}\right)$$

$$= 9.22\ \text{lbf/in}^2 \quad (9.2\ \text{lbf/in}^2)$$

The answer is (A).

Why Other Options Are Wrong

(B) This incorrect answer results when the exponent on the mass flow rate is neglected.

(C) This incorrect answer results when the pressure drop for all operating conditions is incorrectly assumed to be equal to the design pressure drop and therefore independent of the mass flow rate.

(D) This incorrect answer results when the proportionality equation is inverted as

$$\Delta p_{\text{off}} = \Delta p_d\frac{\dot{m}_d^2}{\dot{m}_{\text{off}}^2}$$

SOLUTION 5

A static column of liquid of height h (feet of head) is equivalent to a certain pressure. The relationship between the two can be found from the equation for pressure drop.

Bernoulli Equation

$$p_1 - p_2 = \gamma h_f$$
$$\Delta p = h\gamma_{\text{water}}(\text{SG})_{\text{oil}}$$

The pressure at the discharge for a head of 450 ft of crude oil is

$$p = \frac{(450\ \text{ft})\left(62.4\ \dfrac{\text{lbf}}{\text{ft}^3}\right)(0.86)}{\left(12\ \dfrac{\text{in}}{\text{ft}}\right)^2}$$

$$= 168\ \text{lbf/in}^2 \quad (170\ \text{psi})$$

The answer is (A).

Why Other Options Are Wrong

(B) This incorrect answer results when the specific gravity of crude oil is not included in the calculation of γ.

(C) This incorrect answer results when 1 ft² is converted to 12 in² instead of (12 in)².

(D) This incorrect answer results when pressure is calculated in pounds per square foot, not pounds per square inch.

SOLUTION 6

The P-V diagram for a Carnot cycle is shown. [Common Thermodynamic Cycles]

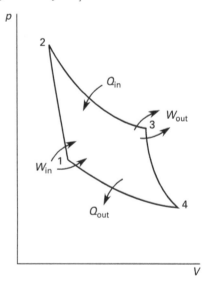

The maximum pressure and temperature occur at state 2, and the minimum pressure and temperature occur at state 4. Therefore, state 2 is where the minimum cylinder volume occurs, and state 4 is where the maximum cylinder volume occurs. The ideal gas law can be used to find the volume at state 2. Then the cycle process can be followed from state 2 to state 3, and then to state 4 where the maximum cylinder volume can be found.

At state 2,

Ideal Gas

$$pV = mRT$$

$$V_2 = \frac{mRT_2}{p_2}$$

$$= \frac{(0.2 \text{ lbm})\left(53.35 \dfrac{\text{ft-lbf}}{\text{lbm-}°\text{R}}\right)(1200°\text{F} + 460°)}{\left(1300 \dfrac{\text{lbf}}{\text{in}^2}\right)\left(12 \dfrac{\text{in}}{\text{ft}}\right)^2}$$

$$= 0.095 \text{ ft}^3$$

The heat added to the air is known, so the volume at state 3 is

$$q_{\text{in}} = p_2 V_2 \ln \frac{V_3}{V_2}$$

$$(5 \text{ Btu})\left(778 \frac{\text{ft-lbf}}{\text{Btu}}\right) = \left(1300 \frac{\text{lbf}}{\text{in}^2}\right)\left(12 \frac{\text{in}}{\text{ft}}\right)^2$$

$$\times (0.095 \text{ ft}^3) \ln \frac{V_3}{0.095 \text{ ft}^3}$$

$$V_3 = 0.12 \text{ ft}^3$$

The process from state 3 to state 4 is reversible and adiabatic, so the maximum cylinder volume is

Carnot Cycle

$$\frac{T_3}{T_1} = \left(\frac{V_1}{V_3}\right)^{k-1}$$

$$\frac{1200°\text{F} + 460°}{80°\text{F} + 460°} = \left(\frac{V_4}{0.12 \text{ ft}^3}\right)^{1.4-1}$$

$$V_4 = 2.0 \text{ ft}^3$$

The answer is (C).

Why Other Options Are Wrong

(A) This incorrect answer results when the volume at state 2 is thought to be the maximum cylinder volume.

(B) This incorrect answer results when the conversion to Rankine is forgotten when calculating V_2.

(D) This incorrect answer results when the volume at state 2 is calculated using the universal gas constant instead of the specific gas constant and the volume is thought to be the maximum cylinder volume.

SOLUTION 7

For a steady-state process, the first law of thermodynamics is

Steady-Flow Systems

$$\sum \dot{m}_i \left(h_i + \frac{\text{v}_i^2}{2} + g_c z_i \right) - \sum \dot{m}_e \left(h_e + \frac{\text{v}_e^2}{2} + g_c z_e \right)$$

$$+ \dot{Q}_{\text{in}} - \dot{W}_{\text{out}} = 0$$

Ignore elevation changes and rearrange to solve for the heat flow rate being added to the compressor.

$$Q_{\text{in}} = \dot{W}_{\text{out}} - \dot{m}\left(h_i - h_e + \frac{\text{v}_i^2}{2g_c} - \frac{\text{v}_e^2}{2g_c} \right)$$

The mass flow rate for the incoming air can be found by combining the continuity equation and the ideal gas equation.

Continuity Equation

$$\dot{m} = \rho A \text{v}$$

Ideal Gas

$$pv = RT$$

$$\dot{m}_i = \frac{A_i \text{v}_i p_i}{R T_i}$$

$$= \frac{(1.0 \text{ ft}^2)\left(20 \dfrac{\text{ft}}{\text{sec}}\right)\left(14.7 \dfrac{\text{lbf}}{\text{in}^2}\right)}{(77°\text{F} + 460°)\left(53.35 \dfrac{\text{ft-lbf}}{\text{lbm-}°\text{R}}\right)}$$

$$\times \left(3600 \frac{\text{sec}}{\text{hr}}\right)\left(12 \frac{\text{in}}{\text{ft}}\right)^2$$

$$= 5320 \text{ lbm/hr}$$

Hydraulic and Fluid Apps.

The heat added to the compressor is

$$Q = (-150 \text{ hp})\left(2545 \frac{\text{Btu}}{\text{hp-hr}}\right) - \left(5320 \frac{\text{lbm}}{\text{hr}}\right)$$

$$\times \begin{pmatrix} 128.34 \dfrac{\text{Btu}}{\text{lbm}} - 196.69 \dfrac{\text{Btu}}{\text{lbm}} \\ + \dfrac{\left(20 \dfrac{\text{ft}}{\text{sec}}\right)^2 - \left(6 \dfrac{\text{ft}}{\text{sec}}\right)^2}{(2)\left(32.17 \dfrac{\text{ft-lbm}}{\text{lbf-sec}^2}\right)} \\ \times \left(\dfrac{1 \text{ Btu}}{778 \text{ ft-lbf}}\right) \end{pmatrix}$$

$$= -18{,}167 \text{ Btu/hr} \quad (-18{,}000 \text{ Btu/hr})$$

The answer is (C).

Why Other Options Are Wrong

(A) This incorrect answer results from omitting the conversion factor from ft-lbf to Btu.

(B) This incorrect answer results from omitting g_c in the equation.

(D) This incorrect answer results from erroneously assuming that vapor compression is an adiabatic process.

SOLUTION 8

The thermal efficiency of a Carnot cycle can be calculated as follows.

Carnot Cycle

$$\eta_{\text{th}} = 1 - \frac{T_L}{T_H}$$

Convert the temperatures to degrees Rankine and calculate the efficiency.

$$\eta_{\text{th}} = 1 - \frac{T_L}{T_H} = 1 - \frac{100°\text{F} + 460°}{300°\text{F} + 460°}$$

$$= 0.26 \quad (26\%)$$

The answer is (A).

Why Other Options Are Wrong

(B) This incorrect answer results when the temperatures are converted using the conversion factor for converting Celsius to kelvin instead of the conversion factor for converting Fahrenheit to Rankine.

(C) This incorrect answer results when neither temperature is converted from Fahrenheit.

(D) This incorrect answer results when the temperatures are converted from Fahrenheit to Celsius instead of Fahrenheit to Rankine.

SOLUTION 9

The thermal efficiency of any heat engine, reversible or irreversible, is given by the equation shown.

Otto Cycle (Gasoline Engine)

$$\eta = 1 - \frac{\dot{Q}_{\text{out}}}{\dot{Q}_{\text{in}}}$$

For reversible engines, the heat transfer ratio can be replaced by the ratio of absolute temperatures, where T_L and T_H are the low and high temperatures, respectively.

Carnot Cycle

$$\eta = 1 - \frac{T_L}{T_H}$$

This shows that both statements I and III are true.

Since this thermal efficiency represents the theoretical best efficiency that can be achieved and the most efficient engine possible is the Carnot cycle, statement II is also true. [Basic Cycles]

Statement IV is false because different reversible engines operating between the same temperature limits will have the same thermal efficiencies.

The answer is (D).

Why Other Options Are Wrong

(A) This is incorrect because statement III is also true.

(B) This is incorrect because statements II and III are also true and statement IV is false.

(C) This is incorrect because statement I is also true.

SOLUTION 10

First, determine the brake horsepower needed for the original pump design. Using the manufacturer's performance curves in the figure, the brake horsepower for 350 gal/min and 200 ft of head is approximately 28 bhp.

Next, determine the brake horsepower for the oversized pump. The oversized pump rating is

$$Q_{\text{os}} = (\text{oversize factor})Q = (1.2)\left(350 \frac{\text{gal}}{\text{min}}\right)$$

$$= 420 \text{ gal/min}$$

$$h_{\text{os}} = (\text{oversize factor})h = (1.2)(200 \text{ ft})$$
$$= 240 \text{ ft}$$

Using the manufacturer's performance curves, the brake horsepower for 420 gal/min and 240 ft of head is approximately 40 bhp.

The larger pump will have to be throttled back to a flow rate of 350 gal/min once put into service. Both pumps deliver 350 gal/min, but the oversized pump needs an additional 12 hp to do it. This 12 hp is wasted energy, so the yearly electrical cost increase is

$$(12 \text{ hp})\left(0.746 \frac{\text{kW}}{\text{hp}}\right)\left(\frac{\$0.07}{\text{kW-hr}}\right)\left(24 \frac{\text{hr}}{\text{day}}\right)\left(365 \frac{\text{day}}{\text{yr}}\right)$$
$$= \$5489 \quad (\$5490)$$

The answer is (D).

Why Other Options Are Wrong

(A) This incorrect answer results when the 20% additional load is not accounted for. Since a 20% increase in pump size would require more power to operate, the cost increase cannot be $0.

(B) This incorrect answer results when the brake horsepower of the original design is misread as 38 bhp.

(C) This incorrect answer results when, in oversizing the pump, the flow rate is increased but the head is not.

SOLUTION 11

Solve the problem using the pump affinity laws for flow and head.

The flow ratio in relation to the speed ratio is

Pump Affinity Laws

$$Q_2 = Q_1\left(\frac{N_2}{N_1}\right)$$
$$\frac{Q_2}{Q_1} = \frac{N_2}{N_1}$$

Substitute the flow rate ratio into the equation for determining the head as it relates to the speed ratio.

Pump Affinity Laws

$$h_2 = h_1\left(\frac{N_2}{N_1}\right)^2$$
$$\frac{h_2}{h_1} = \left(\frac{Q_2}{Q_1}\right)^2$$
$$h_1 = h_2\left(\frac{Q_1}{Q_2}\right)^2$$

With the given curves, the only way to solve this problem is through trial and error. The known values are

$$Q_2 = 550 \text{ gal/min}$$
$$h_2 = 350 \text{ ft}$$
$$N_1 = 3550 \text{ rev/min}$$

Since N_2 is not known, a speed must be guessed. If the resulting values for Q_1 and h_1 lie on the 9.5 in impeller curve, then the speed is correct. If not, a new guess must be tried.

Guess 1: N_2 is 3800 rpm.

$$Q_1 = \left(550 \frac{\text{gal}}{\text{min}}\right)\left(\frac{3550 \frac{\text{rev}}{\text{min}}}{3800 \frac{\text{rev}}{\text{min}}}\right) = 514 \text{ gal/min}$$

$$h_1 = (350 \text{ ft})\left(\frac{3550 \frac{\text{rev}}{\text{min}}}{3800 \frac{\text{rev}}{\text{min}}}\right)^2 = 305 \text{ ft}$$

This point lies near the 10 3/16 in diameter impeller curve, well above the 9.5 in diameter impeller curve.

Guess 2: N_2 is 4100 rpm.

$$Q_1 = \left(550 \frac{\text{gal}}{\text{min}}\right)\left(\frac{3550 \frac{\text{rev}}{\text{min}}}{4100 \frac{\text{rev}}{\text{min}}}\right) = 476 \text{ gal/min}$$

$$h_1 = (350 \text{ ft})\left(\frac{3550 \frac{\text{rev}}{\text{min}}}{4100 \frac{\text{rev}}{\text{min}}}\right)^2 = 262 \text{ ft}$$

This point lies close to the 9.5 in impeller curve, so the speed to produce 550 gal/min and 350 ft of head is approximately 4100 rpm.

Guess 3: N_2 is 4150 rpm.

$$Q_1 = \left(550 \frac{\text{gal}}{\text{min}}\right)\left(\frac{3550 \frac{\text{rev}}{\text{min}}}{4150 \frac{\text{rev}}{\text{min}}}\right) = 470 \text{ gal/min}$$

$$h_1 = (350 \text{ ft})\left(\frac{3550 \frac{\text{rev}}{\text{min}}}{4150 \frac{\text{rev}}{\text{min}}}\right)^2 = 256 \text{ ft}$$

This point lies right on the 9.5 in impeller curve, indicating that the needed speed is 4150 rpm.

The answer is (B).

Why Other Options Are Wrong

(A) This incorrect answer results when, in the calculation of h_1, the N_1 and N_2 speeds in the speed ratio are reversed.

(C) This incorrect answer results when, in the calculation of h_1, the speed ratio is not squared.

(D) This incorrect answer results when, in the calculation of h_1, 550 gal/min is used instead of the 350 ft head.

SOLUTION 12

Use the maximum shear stress failure criterion and a yield stress of 30,000 psi for the steel shaft. [Average Mechanical Properties of Typical Engineering Materials - I-P Units]

The maximum allowable shear stress in the shaft is

$$\tau = 0.5\sigma_y = (0.5)\left(30{,}000 \ \frac{\text{lbf}}{\text{in}^2}\right) = 15{,}000 \text{ psi}$$

The maximum torque is found from the equation

Torsion

$$\tau = \frac{Tr}{J}$$

Rearranging and solving for the torque gives

$$
\begin{aligned}
T &= \frac{\tau J}{r} \\
&= \frac{\left(15{,}000 \ \dfrac{\text{lbf}}{\text{in}^2}\right)\left(\dfrac{\pi}{32}\right)D^4}{1.25 \text{ in}} \\
&= \frac{\left(15{,}000 \ \dfrac{\text{lbf}}{\text{in}^2}\right)\left(\dfrac{\pi}{32}\right)(2.5 \text{ in})^4}{1.25 \text{ in}} \\
&= 46{,}000 \text{ lbf-in}
\end{aligned}
$$

The maximum power that can be transmitted is

$$
\begin{aligned}
P &= T\omega \\
&= (46{,}000 \text{ lbf-in})\left(\frac{1 \text{ ft}}{12 \text{ in}}\right)\left(200 \ \frac{\text{rev}}{\text{min}}\right)\left(2\pi \ \frac{\text{rad}}{\text{rev}}\right) \\
&\quad \times \left(\frac{1 \text{ min}}{60 \text{ sec}}\right)\left(\frac{1 \text{ hp}}{550 \ \dfrac{\text{ft-lbf}}{\text{sec}}}\right) \\
&= 146 \text{ hp} \quad (150 \text{ hp})
\end{aligned}
$$

The answer is (B).

Why Other Options Are Wrong

(A) This incorrect answer results from not converting units of foot-pounds per second to horsepower and adjusting the decimal point to bring the answer into a reasonable range.

(C) This incorrect answer results when the cross-sectional area is used instead of the polar area moment of inertia.

(D) This incorrect answer results when 30,000 psi is used as the maximum shear stress instead of 15,000 psi.

SOLUTION 13

The operating point of a pump is where the system curve crosses the head capacity curve. Because the pump is operating at 3550 rpm, the system curve can be plotted directly on the head capacity curve. The system curve can be found by developing a table of values for Q versus h.

Q_1 (gal/min)	h_1 (ft)
0	0
200	32
400	128
600	288
800	512
1000	800

Plotting these points on the given graph shows that the system curve crosses the head capacity curve at approximately $Q = 800$ gal/min.

The answer is (C).

Why Other Options Are Wrong

(A) This incorrect answer results from switching 200 and 500 in the equation for h_1.

(B) This incorrect answer results from using the point where the pump efficiency curve crosses the pump curve.

(D) This incorrect answer results from neglecting to square the ratio of flow rates.

SOLUTION 14

The cycle thermal efficiency is

Brayton Cycle (Steady-Flow Cycle)

$$\eta_{\text{th}} = \frac{\dot{W}_{\text{net}}}{\dot{Q}_{\text{in}}} = \frac{\dot{W}_{\text{turbine}} - \dot{W}_{\text{pump}}}{\dot{Q}_{\text{in}}}$$

Neglecting pump power, this simplifies to

$$\eta_{\mathrm{th}} = \frac{\dot{W}_{\mathrm{turbine}}}{\dot{Q}_{\mathrm{in}}}$$

If the work done by the turbine and heat output by the condenser occur during the same time interval, the efficiency can be written in terms of work.

$$\eta_{\mathrm{th}} = \frac{W_{\mathrm{turbine}}}{Q_{\mathrm{in}}}$$

The work of the turbine can be found from the turbine output and the mass flow rate.

$$W_{\mathrm{turbine}} = \frac{W_{\mathrm{out}}}{\dot{m}}$$

$$= \frac{(10 \text{ MW})\left(10^{6}\,\dfrac{\text{W}}{\text{MW}}\right)\left(3.413\,\dfrac{\text{Btu}}{\text{hr-W}}\right)}{10^{5}\,\dfrac{\text{lbm}}{\text{hr}}}$$

$$= 341.3 \text{ Btu/lbm}$$

The heat transferred into the system, neglecting the work of the pump, equals the enthalpy of the saturated steam leaving the boiler minus the enthalpy of the saturated liquid leaving the condenser.

$$Q_{\mathrm{in}} = h_{\mathrm{steam}} - h_{\mathrm{liquid}}$$

In the condenser, the condensate is subcooled 6°F. The saturation temperature for water at 2 psia is 126.00°F, so the temperature after subcooling is 120.00°F. [Properties of Saturated Water and Steam (Temperature) - I-P Units]

From steam tables, the enthalpy of this saturated liquid is 88.00 Btu/lbm. [Properties of Saturated Water and Steam (Pressure) - I-P Units]

Also from steam tables, the enthalpy of the saturated steam leaving the boiler at 1000 psia is 1192.6 Btu/lbm. [Properties of Superheated Steam - I-P Units]

Again, neglecting pump work,

$$\eta_{\mathrm{th}} = \frac{W_{\mathrm{turbine}}}{Q_{\mathrm{in}}} = \frac{W_{\mathrm{turbine}}}{h_{\mathrm{steam}} - h_{\mathrm{liquid}}}$$

$$= \frac{341.3\,\dfrac{\text{Btu}}{\text{lbm}}}{1192.6\,\dfrac{\text{Btu}}{\text{lbm}} - 88.0\,\dfrac{\text{Btu}}{\text{lbm}}}$$

$$= 0.31 \quad (31\%)$$

The answer is (C).

Why Other Options Are Wrong

(A) This incorrect answer results from leaving out the conversion from watts to Btu/hr.

(B) This incorrect answer results from adjusting the turbine work by multiplying by the turbine efficiency.

(D) This incorrect answer results from adjusting the turbine work by dividing by the turbine efficiency.

SOLUTION 15

The power generated by the turbine can be found using the following equation.

Turbines

$$\dot{W}_{\mathrm{turbine}} = \dot{m}(h_i - h_e)$$

From steam tables, for superheated steam at 1300 psia and 950°F, the enthalpy is 1467.3 Btu/lbm, and the entropy is 1.599 Btu/lbm-°R. [Properties of Superheated Steam - I-P Units]

$$h_i = 1467.3 \text{ Btu/lbm}$$

Since the expansion is assumed to be isentropic, the entropy at the turbine inlet and exhaust are the same, which can be used to determine the enthalpy at the exhaust.

Using steam tables, find the enthalpy at the exhaust for a pressure of 300 psia and an entropy of 1.599 Btu/lbm-°R. [Properties of Superheated Steam - I-P Units]

$$h_e = 1281.2 \text{ Btu/lbm}$$

No interpolation is needed since the entropy is 1.595 Btu/lbm-°R, which is close enough to the value of s_i. Rearrange the power equation and solve for the mass flow rate, taking into account that the efficiency of the turbine is 75%.

$$\dot{m} = \frac{P_{\mathrm{turbine}}}{\eta(h_i - h_e)}$$

$$= \frac{(9 \times 10^{6} \text{ W})\left(3.413\,\dfrac{\text{Btu}}{\text{W-hr}}\right)}{(0.75)\left(1467.3\,\dfrac{\text{Btu}}{\text{lbm}} - 1281.2\,\dfrac{\text{Btu}}{\text{lbm}}\right)}$$

$$= 220 \times 10^{3} \text{ lbm/hr}$$

The answer is (D).

Why Other Options Are Wrong

(A) This incorrect answer results when the conversion from watts to British thermal units per hour is neglected.

(B) This incorrect answer results when the enthalpy at 1300 psia and 1050°F is used.

(C) This incorrect answer results when the efficiency of the system is ignored.

SOLUTION 16

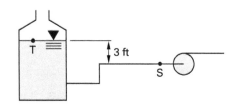

To prevent cavitation, the required net positive suction head (NPSH) must be less than or equal to the available NPSH, where

Centrifugal Pump Characteristics

$$\mathrm{NPSH}_A = h_p + h_z - h_{\mathrm{vpa}} - h_f$$

$$= \frac{p_{\mathrm{atm}}}{\rho g} \pm (z_\mathrm{T} - z_\mathrm{S}) - h_f - \frac{p_{\mathrm{vapor}}}{\rho g}$$

Per pound mass, the equation is

$$\mathrm{NPSH}_A\!\left(\frac{g}{g_c}\right) = \frac{p_{\mathrm{atm}}}{\rho} \pm (z_\mathrm{T} - z_\mathrm{S})\!\left(\frac{g}{g_c}\right) - h_f\!\left(\frac{g}{g_c}\right) - \frac{p_{\mathrm{vapor}}}{\rho}$$

Using points T and S as shown in the diagram, substitute the net positive suction head required (NPSH$_R$ for NPSH$_A$) and solve for h_f.

$$h_f\!\left(\frac{g}{g_c}\right) < (z_\mathrm{T} - z_\mathrm{S})\!\left(\frac{g}{g_c}\right) + \frac{\mathrm{v}_\mathrm{T}^2}{2g_c} + \frac{p_{\mathrm{atm}} - p_{\mathrm{sat}}}{\rho} - \mathrm{NPSH}_R\!\left(\frac{g}{g_c}\right)$$

The velocity of the water in the tank is 0 ft/sec. The water in the vented tank is at atmospheric pressure, and the water level in the tank is 3 ft higher than at the pump intake. From steam tables, the saturation pressure of water at 180°F is 7.52 psia. [Properties of Saturated Water and Steam (Temperature) - I-P Units]

Also from steam tables, the specific volume of water at 180°F is 0.0165 ft³/lbm, which gives a density of 60.6 lbm/ft³. [Properties of Saturated Water and Steam (Pressure) - I-P Units]

$$h_f\!\left(\frac{32.17\ \frac{\mathrm{ft}}{\mathrm{sec}^2}}{32.17\ \frac{\mathrm{ft\text{-}lbm}}{\mathrm{lbf\text{-}sec}^2}}\right) < (3\ \mathrm{ft} - 0\ \mathrm{ft})\!\left(\frac{32.17\ \frac{\mathrm{ft}}{\mathrm{sec}^2}}{32.17\ \frac{\mathrm{ft\text{-}lbm}}{\mathrm{lbf\text{-}sec}^2}}\right) + 0\ \mathrm{ft}$$

$$+ \frac{\left(14.7\ \frac{\mathrm{lbf}}{\mathrm{in}^2} - 7.52\ \frac{\mathrm{lbf}}{\mathrm{in}^2}\right)\!\left(12\ \frac{\mathrm{in}}{\mathrm{ft}}\right)^2}{60.6\ \frac{\mathrm{lbm}}{\mathrm{ft}^3}}$$

$$- (5\ \mathrm{ft})\!\left(\frac{32.17\ \frac{\mathrm{ft}}{\mathrm{sec}^2}}{32.17\ \frac{\mathrm{ft\text{-}lbm}}{\mathrm{lbf\text{-}sec}^2}}\right)$$

$$= 15.06\ \mathrm{ft}$$

Solving for L in the manufacturer's equation,

$$h_f = 1.50 \times 10^{-5} L Q^{1.852}$$

$$L < \frac{h_f}{(1.50 \times 10^{-5}) Q^{1.852}}$$

$$< \frac{15.06\ \mathrm{ft}}{(1.50 \times 10^{-5})\!\left(140\ \frac{\mathrm{gal}}{\mathrm{min}}\right)^{1.852}}$$

$$< 106\ \mathrm{ft} \quad (100\ \mathrm{ft})$$

The answer is (B).

Why Other Options Are Wrong

(A) This incorrect answer results when $z_\mathrm{T} - z_\mathrm{S}$ is left out of the calculations.

(C) This incorrect answer results when the subtraction of NPSHR is neglected.

(D) This incorrect answer results when p_{sat} is left out of the calculations.

SOLUTION 17

When two impellers with different diameters operate at the same speed and efficiency, the affinity laws state that

Pump Affinity Laws

$$Q_2 = Q_1\left(\frac{D_2}{D_1}\right)$$

$$h_2 = h_1\left(\frac{D_2}{D_1}\right)^2$$

$$bhp_2 = bhp_1\left(\frac{D_2}{D_1}\right)^3$$

These laws are only approximate in the case of a clipped impeller, but they are conventionally treated as exact because the difference is usually trivial. Substituting the values in the three equations gives

Pump Affinity Laws

$$Q_2 = Q_1\left(\frac{D_2}{D_1}\right)$$

$$= \left(200\ \frac{gal}{min}\right)\left(\frac{1.5\ ft}{2.0\ ft}\right)$$

$$= 150\ gal/min$$

$$h_2 = h_1\left(\frac{D_2}{D_1}\right)^2$$

$$= (180\ ft)\left(\frac{1.5\ ft}{2.0\ ft}\right)^2$$

$$= 101\ ft \quad (100\ ft)$$

$$bhp_2 = bhp_1\left(\frac{D_2}{D_1}\right)^3$$

$$= (17.0\ hp)\left(\frac{1.5\ ft}{2.0\ ft}\right)^3$$

$$= 7.2\ hp$$

The answer is (D).

Why Other Options Are Wrong

(A) This incorrect answer results when all three ratios are thought to equal D_2/D_1.

(B) This incorrect answer results when the ratio for the diameter is not squared as it relates to the head ratio.

(C) This incorrect answer results when all three ratios are incorrectly thought to equal $(D_2/D_1)^3$.

SOLUTION 18

The linear velocity of the air moving through the fan is

$$v_{air} = pN$$

$$= \left(2.75\ \frac{ft}{rev}\right)\left(490\ \frac{rev}{min}\right)$$

$$= 1347.5\ ft/min$$

The volume of air moving through the fan is

Continuity Equation

$$Q = vA$$

$$= \left(1347.5\ \frac{ft}{min}\right)\left(\frac{\pi}{4}\right)\left(\frac{42\ in}{12\ \frac{in}{ft}}\right)^2$$

$$= 13{,}000\ ft^3/min$$

The answer is (C).

Why Other Options Are Wrong

(A) This incorrect answer results from converting cfm to cfs.

(B) This incorrect answer results from neglecting to square the diameter.

(D) This incorrect answer results from omitting the $\pi/4$ term.

SOLUTION 19

As the flow rate decreases from point A, the compressor performance moves up the pressure-ratio/flow-rate curve to point B. Further reduction in the flow rate causes the compressor to operate at point C, reducing the compressor's pressure capability. As the compressor's performance moves to point C, refrigerant (in the case of refrigeration operation) continues to boil off because of the evaporator's heat load. This builds up the evaporator pressure and decreases the pressure ratio, causing the compressor to shift briefly back to point A, where the cycle repeats. This cycle is known as surging and typically occurs when the compressor is operating at less than 35% of the rated capacity.

The answer is (A).

Why Other Options Are Wrong

(B) Stalling is the breakdown of the airflow through a few stages of a compressor.

(C) Starving is caused by lack of air at the inlet of the compressor.

(D) Throttling is the regulation of flow and head.

SOLUTION 20

The density of water at standard conditions is 62.4 lbm/ft^3. [Properties of Water at Standard Conditions]

From the equation for specific gravity, the density of the kerosene is

Density, Specific Weight, and Specific Gravity

$$SG = \frac{\rho}{\rho_w}$$

$$\rho = (SG)\rho_w$$

$$= (0.85)\left(62.4 \ \frac{lbm}{ft^3}\right)$$

$$= 53 \ lbm/ft^3$$

The flow rate through a Venturi meter is given the equation shown.

Venturi Meters

$$Q = \frac{C_v A_2}{\sqrt{1 - \left(\frac{A_2}{A_1}\right)^2}} \sqrt{2g\left(\frac{p_1}{\gamma} + z_1 - \frac{p_2}{\gamma} - z_2\right)}$$

$C_v = 1.0$ and there is no change in elevation ($z_1 = z_2$). Specific weight, γ, multiplied by g/g_c equals the density, ρ. Solving for the difference in pressure gives

$$Q = \frac{A_2}{\sqrt{1 - \left(\frac{A_2}{A_1}\right)^2}} \sqrt{2g_c\left(\frac{p_1}{\rho} - \frac{p_2}{\rho}\right)}$$

$$p_1 - p_2 = \frac{\rho Q^2\left(1 - \left(\frac{A_2}{A_1}\right)^2\right)}{2g_c A_2^2}$$

$$= \frac{\left(53 \ \frac{lbm}{ft^3}\right)\left(800 \ \frac{gal}{min}\right)^2\left(1 - \left(\frac{0.030 \ ft^2}{0.085 \ ft^2}\right)^2\right)}{(2)\left(32.17 \ \frac{lbm\text{-}ft}{lbf\text{-}sec^2}\right)(0.030 \ ft^2)^2\left(7.48 \ \frac{gal}{ft^3}\right)^2}$$
$$\times \left(60 \ \frac{sec}{min}\right)^2\left(12 \ \frac{in}{ft}\right)^2$$

$$= 17.7 \ lbf/in^2 \quad (17.7 \ psi)$$

The answer is (D).

Why Other Options Are Wrong

(A) This incorrect answer results from using 80 gal/min instead of 800 gal/min for the flow rate.

(B) This incorrect answer results from neglecting to multiply the water density by the kerosene specific gravity.

(C) This incorrect answer results from not squaring the A_2/A_1 ratio.

SOLUTION 21

The mechanical runout is the largest difference between any two readings. The smallest reading is -0.1 mil, and the largest reading is 0.15 mil. The difference is 0.15 mil $- (-0.1 \ mil) = 0.25$ mil.

The answer is (D).

Why Other Options Are Wrong

(A) This incorrect answer is the average of the dial indicator readings, not the runout of the shaft.

(B) This incorrect answer is the largest negative value from zero, not the runout of the shaft.

(C) This incorrect answer is the largest positive value from zero, not the runout of the shaft.

SOLUTION 22

Calculate the velocity of the car. The density of air at standard temperature and pressure is 0.0752 psi. [Properties of Air at Atmospheric Pressure]

Pitot-Static Tubes

$$v = C\sqrt{\frac{2p_w g_c}{\rho}}$$

$$= \left(136.8 \ \frac{sec\text{-}lbf^{1/2}}{ft\text{-}min\text{-}in^{1/2}}\right)$$
$$\times \left(\sqrt{\frac{(2)(7.0 \ in)\left(32.17 \ \frac{lbm\text{-}ft}{lbf\text{-}sec^2}\right)}{0.0752 \ \frac{lbm}{ft^3}}}\right)\left(\frac{60 \ \frac{min}{hr}}{5280 \ \frac{ft}{mi}}\right)$$

$$= 120 \ mi/hr$$

The answer is (C).

Why Other Options Are Wrong

(A) This incorrect answer results from neglecting the minutes-to-hour conversion.

(B) This incorrect answer results from forgetting the 7.0 factor in the radical.

(D) This incorrect answer results from transposing the 3 and 6 in the 136.8 term.

SOLUTION 23

The pump efficiency can be estimated by comparing the motor's power output to the pump's power consumption.

Using the motor efficiency, η_m, find the power output by the motor.

Mechanical Power

$$\eta_m P_{\text{elec}} = P_{\text{mech}}$$
$$P_{\text{mech}} = \eta_m P_{\text{elec}}$$
$$= \frac{\eta_m \sqrt{3}\, IV(\text{pf})}{1000}$$
$$= \frac{(0.93)(\sqrt{3})(100 \text{ A})(2300 \text{ V})(0.87)}{1000}$$
$$= 322 \text{ kW}$$

Convert the power output of the pump to water horsepower using the pressure differential given in the problem statement.

Pump Power Equation

$$\text{whp} = \frac{Q \Delta p}{1714}$$

$$P_{p,\text{out}} = \frac{\left(500 \, \dfrac{\text{gal}}{\text{min}}\right)\left(850 \, \dfrac{\text{lbf}}{\text{in}^2}\right)\left(745.7 \, \dfrac{\text{W}}{\text{hp}}\right)}{\left(1714 \, \dfrac{\text{gpm-lbf}}{\text{in}^2\text{-hp}}\right)\left(1000 \, \dfrac{\text{W}}{\text{kW}}\right)}$$
$$= 185 \text{ kW}$$

Equating the output power from the motor with the input power needed by the pump,

$$P_{p,\text{in}} = P_{\text{mech}} = 322 \text{ kW}$$

Solving for the approximate pump efficiency gives

$$\eta_p = \frac{P_{p,\text{out}}}{P_{p,\text{in}}}$$

$$P_{p,\text{in}} = \frac{P_{p,\text{out}}}{\eta_p}$$
$$= \frac{185 \text{ kW}}{322 \text{ kW}}$$
$$= 0.57 \quad (57\%)$$

The answer is (A).

Why Other Options Are Wrong

(B) This incorrect answer results from neglecting the conversion from horsepower to kilowatts.

(C) This incorrect answer results from neglecting to use the power factor and the square root of 3.

(D) This incorrect answer results from neglecting to use the square root of 3.

SOLUTION 24

The diagram shows the relevant temperature points in the system.

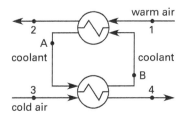

Use the equations for the sensible effectiveness and maximum sensible heat transfer rate of a heat-recovery ventilator to find the change in air temperature.

Heat-Recovery Ventilator (HRV)—Sensible Energy Recovery

$$\dot{q}_{s,\text{max}} = 60 C_{\text{min}} (T_3 - T_1)$$

$$\frac{\dot{q}_s}{\dot{q}_{s,\text{max}}} = \frac{\dot{m}_s c_{\text{ps}} (T_2 - T_1)}{C_{\text{min}} (T_3 - T_1)}$$

$$T_2 - T_1 = \frac{\dot{q}_s C_{\text{min}} (T_3 - T_1)}{\dot{q}_{s,\text{max}} \dot{m}_s c_{\text{ps}}}$$
$$= \frac{\dot{q}_s C_{\text{min}} (T_3 - T_1)}{\left(60 C_{\text{min}} (T_3 - T_1)\right) \dot{m}_s c_{\text{ps}}}$$
$$= \frac{\dot{q}_s}{60 \dot{m}_s c_{\text{ps}}}$$
$$= \frac{200{,}000 \, \dfrac{\text{Btu}}{\text{hr}}}{\left(60 \, \dfrac{\text{min}}{\text{hr}}\right)\left(600 \, \dfrac{\text{lbm}}{\text{min}}\right)\left(0.24 \, \dfrac{\text{Btu}}{\text{lbm-}^\circ\text{F}}\right)}$$
$$= 23.1^\circ\text{F}$$

The changes in coolant temperature are similarly equal and opposite. For the same heat transfer rate,

$$T_A - T_B = \frac{Q}{(\dot{m} c_p)_{\text{coolant}}}$$
$$= \frac{200{,}000 \, \dfrac{\text{Btu}}{\text{hr}}}{\left(60 \, \dfrac{\text{min}}{\text{hr}}\right)\left(250 \, \dfrac{\text{lbm}}{\text{min}}\right)\left(1 \, \dfrac{\text{Btu}}{\text{lbm-}^\circ\text{F}}\right)}$$
$$= 13.3^\circ\text{F}$$

The average coolant temperature is

$$\overline{T}_{coolant} = \frac{T_A + T_B}{2} = \frac{T_1 + T_3}{2}$$
$$= \frac{10°F + 75°F}{2}$$
$$= 42.5°F$$

Solve for the left and right coolant temperatures, T_A and T_B.

$$T_A = \overline{T}_{coolant} + \frac{T_A - T_B}{2}$$
$$= 42.5°F + \frac{13.3°F}{2}$$
$$= 49.2°F \quad (49°F)$$
$$T_B = \overline{T}_{coolant} - \frac{T_A - T_B}{2}$$
$$= 42.5°F - \frac{13.3°F}{2}$$
$$= 35.9°F \quad (36°F)$$

The answer is (B).

Why Other Options Are Wrong

(A) This incorrect answer results when $(T_2 - T_1)$ is used instead of $(T_A - T_B)$ and T_A and T_B are interchanged.

(C) This incorrect answer results when $(T_2 - T_1)$ is used instead of instead of $(T_A - T_B)$ to solve for the left coolant temperature.

(D) This incorrect answer results when it is assumed that $T_B = T_1$ and $T_A = T_3$.

SOLUTION 25

The mass flow rate at points A and B must be equal.

$$\dot{m}_A = \dot{m}_B$$

From the continuity equation,

Continuity Equation

$$\dot{m} = \rho A v$$

$$\rho_A A_A v_A = \rho_B A_B v_B$$

The cross-sectional areas at points A and B are equal, but the densities at A and B are not equal because CO_2 is compressible. The densities at A and B can be calculated using the ideal gas law.

Ideal Gas

$$pV = mRT$$
$$p = \frac{m}{V}RT$$
$$\rho = \frac{p}{RT}$$

The specific gas constant, R, for CO_2 is 35.11 ft-lbf/lbm-°R. [Thermal and Physical Properties of Ideal Gases (at Room Temperature)]

Calculate the densities at points A and B.

$$\rho_A = \frac{p_A}{RT_A}$$
$$= \frac{\left(40 \frac{lbf}{in^2} + 14.7 \frac{lbf}{in^2}\right)\left(12 \frac{in}{ft}\right)^2}{\left(35.11 \frac{ft\text{-}lbf}{lbm\text{-}°R}\right)(70°F + 460°)}$$
$$= 0.423 \ lbm/ft^3$$

$$\rho_B = \frac{p_B}{RT_B}$$
$$= \frac{\left(30 \frac{lbf}{in^2} + 14.7 \frac{lbf}{in^2}\right)\left(12 \frac{in}{ft}\right)^2}{\left(35.11 \frac{ft\text{-}lbf}{lbm\text{-}°R}\right)(90°F + 460°)}$$
$$= 0.333 \ lbm/ft^3$$

Determine the velocity at point B.

$$\rho_A v_A = \rho_B v_B$$
$$v_B = \frac{\rho_A v_A}{\rho_B}$$
$$= \frac{\left(0.423 \frac{lbm}{ft^3}\right)\left(20 \frac{ft}{sec}\right)}{0.333 \frac{lbm}{ft^3}}$$
$$= 25.4 \ ft/sec$$

Using the continuity equation, solve for the volumetric flow rate at point B.

Continuity Equation

$$Q_B = A v_B$$

$$= \frac{\pi}{4}\left(\frac{6 \text{ in}}{12 \frac{\text{in}}{\text{ft}}}\right)^2 \left(25.4 \frac{\text{ft}}{\text{sec}}\right)\left(60 \frac{\text{sec}}{\text{min}}\right)$$

$$= 299.2 \text{ ft}^3/\text{min} \quad (300 \text{ ft}^3/\text{min})$$

The answer is (C).

Why Other Options Are Wrong

(A) This incorrect answer uses cfs instead of cfm.

(B) This incorrect answer is the flow rate at point A, not point B.

(D) This incorrect answer results when gage pressure is used instead of absolute pressure.

SOLUTION 26

The log mean temperature difference (LMTD) is

Log Mean Temperature Difference (LMTD)

$$\Delta T_{lm} = \frac{(T_{Ho} - T_{Ci}) - (T_{Hi} - T_{Co})}{\ln\left(\frac{T_{Ho} - T_{Ci}}{T_{Hi} - T_{Co}}\right)}$$

$$= \frac{(215°\text{F} - 90°\text{F}) - (1300°\text{F} - 212°\text{F})}{\ln\left(\frac{215°\text{F} - 90°\text{F}}{1300°\text{F} - 212°\text{F}}\right)}$$

$$= 445.55°\text{F} \quad (446°\text{F})$$

The answer is (C).

Why Other Options Are Wrong

(A) This incorrect answer results when the natural log is used instead of the log base 10.

(C) This incorrect answer results when the factor of 2.3 in the denominator is neglected.

(D) This incorrect answer results when the average of the entering and exit temperatures of the gas is taken.

SOLUTION 27

The humidity ratio is the ratio of the mass of water vapor to the mass of dry air.

Psychrometric Properties

$$W = \frac{m_w}{m_{da}}$$

Rearrange the ideal gas law to solve for the mass.

Ideal Gas

$$pV = mRT$$

$$m = \frac{pV}{RT}$$

Combining the equation for the humidity ratio with the rearranged ideal gas law gives

$$W = \frac{m_w}{m_{da}} = \frac{\dfrac{p_w V_w}{R_w T_w}}{\dfrac{p_{da} V_{da}}{R_{da} T_{da}}}$$

The volumes and temperatures are equal, so the equation simplifies to

$$W = \frac{\dfrac{p_w}{R_w}}{\dfrac{p_{da}}{R_{da}}}$$

Find the actual partial air pressure.

Psychrometric Properties

$$p_{da} = p_{total} - p_w = 30 \frac{\text{lbf}}{\text{in}^2} - 2.79 \frac{\text{lbf}}{\text{in}^2} = 27.21 \text{ lbf/in}^2$$

The specific gas constant of air is 53.35 ft-lbf/lbm-°R, and the specific gas constant of steam is 85.78 ft-lbf/lbm-°R. [Thermal and Physical Properties of Ideal Gases (at Room Temperature)]

Solve for the humidity ratio.

$$W = \frac{\dfrac{p_w}{R_w}}{\dfrac{p_{da}}{R_{da}}} = \frac{p_w R_{da}}{p_{da} R_w}$$

$$= \frac{\left(2.79 \dfrac{\text{lbf}}{\text{in}^2}\right)\left(53.35 \dfrac{\text{ft-lbf}}{\text{lbm-°R}}\right)}{\left(27.21 \dfrac{\text{lbf}}{\text{in}^2}\right)\left(85.78 \dfrac{\text{ft-lbf}}{\text{lbm-°R}}\right)}$$

$$= 0.064 \text{ lbm}_{vapor}/\text{lbm}_{air}$$

The answer is (A).

Why Other Options Are Wrong

(B) This incorrect answer results when the saturation pressure is used instead of the partial vapor pressure.

(C) This incorrect answer results when the two specific gas constants in the relative humidity equation are reversed from what they should be.

(D) This incorrect answer results when the pressures in the relative humidity equation are converted to gauge pressures.

SOLUTION 28

Because the compressor is not cooled, the power needed for isentropic compression of an ideal gas from the inlet pressure, p_i, to the discharge pressure, p_e, is the theoretical adiabatic power.

Compressors

$$\dot{W}_{\text{comp}} = \frac{\dot{m}p_i k}{(k-1)\rho_i \eta_c}\left[\left(\frac{p_e}{p_i}\right)^{1-\frac{1}{k}} - 1\right]$$

Use the continuity equation to rewrite the equation as follows.

Continuity Equation

$$\dot{m} = Q\rho$$

$$\dot{W}_{\text{comp}} = \frac{(Q_i\rho_i)p_i k}{(k-1)\rho_i \eta_c}\left[\left(\frac{p_e}{p_i}\right)^{1-\frac{1}{k}} - 1\right]$$

$$= \frac{Q_i p_i k}{(k-1)\eta_c}\left[\left(\frac{p_e}{p_i}\right)^{1-\frac{1}{k}} - 1\right]$$

Find the adiabatic power needed. The ratio of specific heats is 1.4 for air. [Mach Number]

The inlet pressure for the atmospheric air is 14.7 psia.

$$\dot{W}_{\text{comp}} = \frac{Q_i p_i k}{(k-1)\eta_c}\left[\left(\frac{p_e}{p_i}\right)^{1-\frac{1}{k}} - 1\right]$$

$$= \left[\frac{\left(50{,}000\ \frac{\text{ft}^3}{\text{min}}\right)\left(14.7\ \frac{\text{lbf}}{\text{in}^2}\right)\left(12\ \frac{\text{in}}{\text{ft}}\right)^2(1.4)}{(1.4-1)(0.79)\left(60\ \frac{\text{sec}}{\text{min}}\right)\left(550\ \frac{\text{lbf-ft}}{\text{hp-sec}}\right)}\right]$$

$$\times\left[\left(\frac{53\ \frac{\text{lbf}}{\text{in}^2}}{14.7\ \frac{\text{lbf}}{\text{in}^2}}\right)^{1-\frac{1}{1.4}} - 1\right]$$

$$= 6288\ \text{hp} \quad (6300\ \text{hp})$$

The answer is (D).

Why Other Options Are Wrong

(A) This incorrect answer results when the flow is converted from cfm to cfs by dividing by 60^2 instead of 60.

(B) This incorrect answer results when the specific heats term at the beginning of the adiabatic power equation is ignored.

(C) This incorrect answer results when 35 psia is used as the discharge pressure instead of 53 psia.

SOLUTION 29

The head loss due to friction is given by

Head Loss Due to Flow: Darcy-Weisbach Equation

$$h_f = f\frac{L}{D}\frac{\text{v}^2}{2g}$$

The minor losses in pipe fittings, contractions, and expansions is given by

Minor Losses in Pipe Fittings, Contractions, and Expansions

$$h_{f,\text{fitting}} = k\frac{\text{v}^2}{2g}$$

The value for k is the loss factor for entrances, exits, fittings, and valves. The equation for minor losses becomes

$$h_{f,\text{fitting}} = \left(k_{\text{exit}}\frac{\text{v}^2}{2g} + 2k_{\text{elbow}}\frac{\text{v}^2}{2g} + k_{\text{entrance}}\frac{\text{v}^2}{2g}\right)$$

Combining the equations for head loss due to friction and minor losses gives the equation for total head loss, h_L.

$$h_L = \left(f \frac{L}{D} \frac{\text{v}^2}{2g} + k_{\text{exit}} \frac{\text{v}^2}{2g} + 2k_{\text{elbow}} \frac{\text{v}^2}{2g} + k_{\text{entrance}} \frac{\text{v}^2}{2g} \right)$$

$$= \frac{\text{v}^2}{2g} \left(f \frac{L}{D} + k_{\text{exit}} + 2k_{\text{elbow}} + k_{\text{entrance}} \right)$$

From tables of k-factors for pipe fittings, k_{exit} is 1.0, k_{elbow} is 0.29, and k_{entrance} is 0.5. [K-Factors—Threaded Pipe Fittings]

The average water velocity in the pipe is found using the continuity equation.

Continuity Equation

$$Q = A\text{v}$$

$$\text{v} = \frac{Q}{A}$$

$$= \frac{\left(60 \frac{\text{ft}^3}{\text{min}} \right) \left(\frac{1 \text{ min}}{60 \text{ sec}} \right)}{\left(\frac{\pi}{4} \right) \left((6 \text{ in}) \left(\frac{1 \text{ ft}}{12 \text{ in}} \right) \right)^2}$$

$$= 5.1 \text{ ft/sec}$$

The total head loss is

$$h_L = \frac{\text{v}^2}{2g} \left(f \frac{L}{D} + k_{\text{exit}} + 2k_{\text{elbow}} + k_{\text{entrance}} \right)$$

$$= \frac{\left(5.1 \frac{\text{ft}}{\text{sec}} \right)^2}{(2) \left(32.17 \frac{\text{ft}}{\text{sec}^2} \right)}$$

$$\left((0.03) \times \left(\frac{(50 \text{ ft}) \left(12 \frac{\text{in}}{\text{ft}} \right)}{6 \text{ in}} \right) + 1.0 + (2)(0.29) + 0.5 \right)$$

$$= 2.05 \text{ ft} \quad (2.1 \text{ ft})$$

The answer is (D).

Why Other Options Are Wrong

(A) This incorrect answer results when the head loss due to friction is neglected.

(B) This incorrect answer results when head losses due to the exit and entrance are neglected.

(C) This incorrect answer results when head losses due to the exit, 90° elbows, and entrance are neglected.

SOLUTION 30

Pump inlet pressure required is given as 20 psi, and the density of water at 68°F is 62.4 lbm/ft³. Convert the inlet pressure to feet of head at the given water temperature. [Properties of Water at Standard Conditions]

Units

$$p = \frac{\rho g h}{g_c}$$

$$h = \frac{p g_c}{\rho g}$$

$$= \frac{\left(20 \frac{\text{lbf}}{\text{in}^2} \right) \left(12 \frac{\text{in}}{\text{ft}} \right)^2 \left(32.17 \frac{\text{lbm-ft}}{\text{lbf-sec}^2} \right)}{\left(62.4 \frac{\text{lbm}}{\text{ft}^3} \right) \left(32.17 \frac{\text{ft}}{\text{sec}^2} \right)}$$

$$= 46.2 \text{ ft} \quad (46 \text{ ft})$$

The answer is (C).

Why Other Options Are Wrong

(A) This incorrect answer results when the inches-to-feet conversion is not squared.

(B) This incorrect answer results when the net positive suction head of 20 psi is used as the solution.

(D) This incorrect answer results when the atmospheric pressure of 14.7 psi is added to the 20 psi.

SOLUTION 31

Find the electrical output power in kilowatts.

Motor Phases

$$P_{\text{electrical}} = \frac{\sqrt{3} \, I V \, (\text{pf})}{1000}$$

$$= \frac{\sqrt{3} \, (20 \text{ A})(240 \text{ V}) \, (0.8)}{1000}$$

$$= 6.65 \text{ kW}$$

Use the generator's efficiency to find the mechanical input power to the generator.

Electrical Power

$$\eta_G P_{\text{mechanical}} = P_{\text{electrical}}$$

$$P_{\text{mechanical}} = \frac{P_{\text{electrical}}}{\eta_G}$$

$$= \frac{6.65 \text{ kW}}{0.90}$$

$$= 7.4 \text{ kW}$$

The torque can be found using the following equation.

$$P = 2\pi T\omega \quad [\omega \text{ in rev/sec}]$$

$$T = \frac{P}{2\pi\omega}$$

Find the torque needed to drive the generator at a rotational speed of 1800 rpm. 1 ft-lbf is equal to 1.356 W-sec. [Measurement Relationships]

$$T = \frac{P_{\text{mechanical}}}{2\pi\omega}$$

$$= \frac{\left(\dfrac{7.4 \text{ kW}}{1.356 \frac{\text{W-sec}}{\text{ft-lbf}}}\right)\left(1000 \frac{\text{W}}{\text{kW}}\right)}{\left(\dfrac{\left(2\pi \frac{\text{rad}}{\text{rev}}\right)\left(1800 \frac{\text{rev}}{\text{min}}\right)}{60 \frac{\text{sec}}{\text{min}}}\right)}$$

$$= 28.9 \text{ lbf-ft} \quad (29 \text{ lbf-ft})$$

The answer is (B).

Why Other Options Are Wrong

(A) This incorrect answer results when power for a single-phase generator is used when calculating $p_{\text{electrical}}$.

(C) This incorrect answer results when the power factor and efficiency numbers are interchanged.

(D) This incorrect answer results when $P_{\text{electrical}}$ is calculated using 3 instead of the square root of 3.

SOLUTION 32

The lowest temperature that can be achieved using evaporative cooling occurs when the air leaves the cooler in its saturated state. Since evaporative cooling is almost identical to the adiabatic saturation process, the process follows lines of constant wet-bulb temperature on the psychrometric chart. From the psychrometric chart, the minimum temperature is found to be approximately 71°F. [ASHRAE Psychrometric Chart No. 1 - Normal Temperature at Sea Level]

The answer is (C).

Why Other Options Are Wrong

(A) This incorrect answer results when the enthalpy value is used.

(B) This incorrect answer results when a horizontal line on the psychrometric chart is followed to the left until it intersects the saturation line.

(D) This incorrect answer results when the 30% relative humidity line is accidently used.

SOLUTION 33

Find the pressure head by rearranging the equation for fluid pressure. [Properties of Water at Standard Conditions]

Units

$$p = \frac{\rho g h}{g_c}$$

$$h = \frac{p g_c}{\rho g}$$

$$= \frac{p g_c}{\rho(\text{SG})g}$$

$$= \frac{\left(420 \frac{\text{lbf}}{\text{in}^2}\right)\left(12 \frac{\text{in}}{\text{ft}}\right)^2\left(32.17 \frac{\text{lbm-ft}}{\text{lbf-sec}^2}\right)}{\left(62.4 \frac{\text{lbm}}{\text{ft}^3}\right)(2.0)\left(32.17 \frac{\text{ft}}{\text{sec}^2}\right)}$$

$$= 485 \text{ ft}$$

Find the mass flow rate of the concrete. Be sure to account for the specific gravity of the concrete.

Continuity Equation

$$\dot{m} = \rho Q$$

$$\dot{m} = \rho(\text{SG})Q$$

$$= \left(62.4 \frac{\text{lbm}}{\text{ft}^3}\right)(2.0)\left(1.20 \frac{\text{ft}^3}{\text{sec}}\right)$$

$$= 149.76 \text{ lbm/sec}$$

The water horsepower is

Pump Power Equation

$$\text{whp} = \frac{\dot{m}\,\Delta h}{33,000}$$

$$= \frac{\left(149.76 \frac{\text{lbm}}{\text{sec}}\right)\left(60 \frac{\text{sec}}{\text{min}}\right)(485 \text{ ft})}{33,000 \frac{\text{ft-lbm}}{\text{min-hp}}}$$

$$= 132 \text{ hp} \quad (130 \text{ hp})$$

The answer is (C).

Why Other Options Are Wrong

(A) This incorrect answer results when the specific gravity is neglected in the whp equation.

(B) This incorrect answer results when the flow rate of 1.2 ft³/sec is neglected in the whp equation.

(D) This incorrect answer results when 2.1 instead of 1.2 is used for the flow rate in the whp equation.

SOLUTION 34

For the airplane to fly straight and level, the lift force, F_L, must be equal to the weight of the airplane.

$$W = F_L$$

Use the drag force equation to calculate the lift coefficient.

Drag Force

$$F_D = \frac{C_D \rho \mathrm{v}^2 A}{2}$$

$$F_L = \frac{mg}{g_c} = \frac{C_L \rho \mathrm{v}^2 A}{2 g_c}$$

$$mg = \frac{C_L \rho \mathrm{v}^2 A}{2}$$

From a table of temperature and altitude corrections for air, the air density at 10,000 ft is 0.0515 lbm/ft^3. [Temperature and Altitude Corrections for Air]

Solving for the lift coefficient gives

$$mg = \frac{C_L \rho \mathrm{v}^2 A}{2}$$

$$C_L = \frac{2mg}{\rho \mathrm{v}^2 A}$$

$$= \frac{(2)(2450 \text{ lbm})\left(32.17 \dfrac{\text{ft}}{\text{sec}}\right)}{\left(0.0515 \dfrac{\text{lbm}}{\text{ft}^3}\right)}$$
$$\times \left(\left(140 \dfrac{\text{mi}}{\text{hr}}\right)\left(5280 \dfrac{\text{ft}}{\text{mi}}\right)\left(\dfrac{1 \text{ hr}}{3600 \text{ sec}}\right)\right)^2$$
$$\times \left(174 \text{ ft}^2\right)$$

$$= 0.42$$

Determine the angle of attack of the wings using interpolation.

$$\frac{0.75 - 0.25}{5° - 0°} = \frac{0.42 - 0.25}{\alpha - 0°}$$

$$\alpha = 1.7°$$

The answer is (C).

Why Other Options Are Wrong

(A) This incorrect answer results when the factor of g is neglected in the equation for C_L.

(B) This incorrect answer results when the lift coefficient is used for the angle of attack.

(D) This incorrect answer results when 4250 lbm is mistakenly used for the airplane mass.

SOLUTION 35

As shown in a free-body diagram of the sphere, the forces acting on the sphere are the buoyancy force, F_B, the weight, W, and the drag force, F_D.

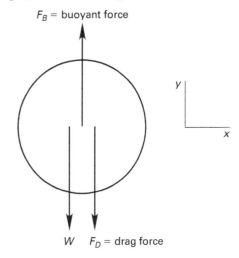

Because the sphere is rising at a constant velocity (acceleration = 0), the equation of motion is

$$\sum F_y = ma_y$$

$$F_B - W - F_D = 0$$

Calculate the volume of the sphere.

Sphere

$$V = \frac{4\pi r^3}{3}$$

$$= \left(\frac{(4)(\pi)}{3}\right)\left(\frac{3 \text{ in}}{12 \dfrac{\text{in}}{\text{ft}}}\right)^3$$

$$= 0.065 \text{ ft}^3$$

Find the buoyancy force using the equation for a sphere.

$$F_B = \frac{\rho_{\text{oil}} V g}{g_c} = \frac{\rho_{\text{oil}}\left(\dfrac{4\pi r^3}{3}\right) g}{g_c}$$

$$= \frac{\left(56 \dfrac{\text{lbm}}{\text{ft}^3}\right)(0.065 \text{ ft}^3)\left(32.17 \dfrac{\text{ft}}{\text{sec}^2}\right)}{32.17 \dfrac{\text{lbm-ft}}{\text{lbf-sec}^2}}$$

$$= 3.64 \text{ lbf}$$

The weight is

$$W = \frac{mg}{g_c} = \frac{(2.5 \text{ lbm})\left(32.17 \; \dfrac{\text{ft}}{\text{sec}^2}\right)}{32.17 \; \dfrac{\text{lbm-ft}}{\text{lbf-sec}^2}} = 2.5 \text{ lbf}$$

The drag force is

Drag Force

$$F_D = \frac{C_D \rho_{\text{oil}} v^2 A}{2}$$

$$= \frac{C_D \rho_{\text{oil}} v^2 \left(\dfrac{\pi D^2}{4}\right)}{2 g_c}$$

$$= \frac{(0.5)\left(56 \; \dfrac{\text{lbm}}{\text{ft}^3}\right)(v^2)\dfrac{\pi}{4}\left(\dfrac{6 \text{ in}}{12 \; \dfrac{\text{in}}{\text{ft}}}\right)^2}{(2)\left(32.17 \; \dfrac{\text{lbm-ft}}{\text{lbf-sec}^2}\right)}$$

$$= 0.0854 v^2 \text{ lbf-sec}^2/\text{ft}^2$$

Substituting into the equation of motion gives

$$F_B - W - F_D = 0$$
$$3.64 \text{ lbf} - 2.5 \text{ lbf}$$
$$-0.0854 v^2 \; \frac{\text{lbf-sec}^2}{\text{ft}^2} = 0$$

$$v = \sqrt{\frac{3.64 \text{ lbf} - 2.5 \text{ lbf}}{0.0854 \; \dfrac{\text{lbf-sec}^2}{\text{ft}^2}}}$$

$$= 3.65 \text{ ft/sec} \quad (3.7 \text{ ft/sec})$$

The answer is (B).

Why Other Options Are Wrong

(A) This incorrect answer results when the acceleration of the sphere is calculated using only the buoyancy force.

(C) This incorrect answer results when the weight is neglected.

(D) This incorrect answer results when the area instead of the volume is used when calculating the buoyancy force.

SOLUTION 36

Let point 1 be the surface of the canal and point 2 be a point at the valve. The dimensions of the lock and canal can be represented as shown.

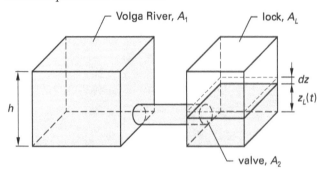

Use the Bernoulli equation.

Bernoulli Equation

$$\frac{p_1}{\rho g} + z_1 + \frac{v_1^2}{2g} = \frac{p_2}{\rho g} + z_2 + \frac{v_2^2}{2g} + h_f$$

$$\frac{p_1}{\rho} + \frac{v_1^2}{2g_c} + z_1 \frac{g}{g_c} = \frac{p_2}{\rho} + \frac{v_2^2}{2g_c} + z_2 \frac{g}{g_c} + h_{\text{losses}} \frac{g}{g_c}$$

From the illustration, the static pressure head, p_2, can be found as shown.

$$p_2 = \rho \frac{g}{g_c} z_L(t)$$

The surface area of the lock is much larger than the surface area of the opening into the lock from the valve, so v_1 is negligible. From the illustration, z_1 is equivalent to h, and z_2 and p_1 are both equal to 0. Neglecting losses and defining $z(t) \equiv z_L(t)$, the Bernoulli equation can be simplified to solve for v_2 as shown.

$$v_2 = \sqrt{2g(h - z(t))}$$

From the continuity equation, the equation for the water flow through the valve is

Continuity Equation

$$Q = Av$$
$$Q_2 = A_2 v_2 = A_2 \sqrt{2g(h - z(t))}$$

The equation for the flow rate into the lock is

$$Q_L = v_L A_L$$

Q_2, and Q_L are equal, so the equation for the rate at which the water in the lock rises is

$$v_L = \frac{Q}{A_L} = \frac{A_2}{A_L}\left(\sqrt{2g(h-z(t))}\right)$$

When the lock is half-full, the water height in the lock is 18 ft. The velocity at that time is

$$v_L = \left(\frac{140 \text{ ft}^2}{(950 \text{ ft})(100 \text{ ft})}\right)\left(\sqrt{(2)\left(32.17\frac{\text{ft}}{\text{sec}^2}\right)(36 \text{ ft} - 18 \text{ ft})}\right)$$

$$= \left(0.05\frac{\text{ft}}{\text{sec}}\right)\left(60\frac{\text{sec}}{\text{min}}\right)$$

$$= 3 \text{ ft/min} \quad (3.0 \text{ ft/min})$$

The answer is (B).

Why Other Options Are Wrong

(A) This incorrect answer is the value in in/sec without converting.

(C) This incorrect answer results when 10 ft is used for the dimensions of the lock instead of 100 ft.

(D) This incorrect answer results if the square root is not taken.

SOLUTION 37

The change in pressure in each line can be calculated by using the Bernoulli equation and including a term for energy loss due to friction.

Duct Design: Bernoulli Equation

$$\frac{v^2}{2g_c} + \frac{p}{\rho} + \frac{gz}{g_c} = \text{constant}$$

$$\frac{p_1}{\rho} + \frac{v_1^2}{2g_c} + \frac{z_1 g}{g_c} = \frac{p_2}{\rho} + \frac{v_2^2}{2g_c} + \frac{z_2 g}{g_c} + \frac{h_f g}{g_c}$$

In each line, $v_1 = v_2$ and $z_1 = z_2$. Simplifying the equation gives

$$\frac{p_1}{\rho} = \frac{p_2}{\rho} + \frac{h_f g}{g_c}$$

Rearranging in terms of friction loss gives

Bernoulli Equation

$$p_1 - p_2 = \rho g h_f$$

$$\Delta p = \frac{\rho g h_f}{g_c}$$

$$h_f = \frac{\Delta p g_c}{\rho g}$$

The head loss due to friction is given by the Darcy-Weisbach equation.

Head Loss Due to Flow: Darcy-Weisbach Equation

$$h_f = f\left(\frac{L}{D}\right)\left(\frac{v^2}{2g}\right)$$

Combine the equations, adding the conversion from pounds force to pounds mass, and rearranging to solve for the change in pressure.

$$\Delta p = f\left(\frac{L}{D}\right)\left(\frac{v^2}{2g}\right)\left(\frac{\rho g}{g_c}\right)$$

The pressure drop per unit length can be expressed as

$$\frac{\Delta p}{L} = \frac{f \rho v^2}{2 D g_c}$$

The friction factor, f, and the density of oil, ρ, are the same for the main line and the feeder lines. Velocity can be expressed as

Continuity Equation

$$Q = Av$$

$$v = \frac{Q}{A} = \frac{Q}{\frac{\pi D^2}{4}} = \frac{4Q}{\pi D^2}$$

If Q is the flow rate in each of the feeder lines, then the flow rate in the main line is $3Q$. The ratio of the pressure drop per unit length in the main line to that in the feeder lines is

$$\frac{\left(\dfrac{\Delta p}{L}\right)_{\text{main}}}{\left(\dfrac{\Delta p}{L}\right)_{\text{feed}}} = \frac{\dfrac{f\rho v^2_{\text{main}}}{2D_{\text{main}}g_c}}{\dfrac{f\rho v^2_{\text{feed}}}{2D_{\text{feed}}g_c}} = \frac{\dfrac{v^2_{\text{main}}}{D_{\text{main}}}}{\dfrac{v^2_{\text{feed}}}{D_{\text{feed}}}}$$

$$= \frac{\dfrac{\left(\dfrac{4(3Q)}{\pi D^2_{\text{main}}}\right)^2}{D_{\text{main}}}}{\dfrac{\left(\dfrac{4Q}{\pi D^2_{\text{feed}}}\right)^2}{D_{\text{feed}}}}$$

$$= \frac{9D^5_{\text{feed}}}{D^5_{\text{main}}}$$

$$= \frac{(9)(15\text{ in})^5}{(29\text{ in})^5}$$

$$= 0.3332 \quad (0.33)$$

The answer is (B).

Why Other Options Are Wrong

(A) This incorrect answer results when the flow in the main line is taken as Q instead of $3Q$.

(C) This incorrect answer results when the velocities are not squared.

(D) This incorrect answer is the ratio of the feeder line to the main line.

3 Energy/Power System Applications

PROBLEM 1

The number of independent quantities needed to specify the thermodynamic state of a system in equilibrium is equal to

(A) three

(B) the number of phases of the substance present in the system

(C) the number of possible phases a substance can have

(D) the number of possible work modes plus one

Hint: The number of independent quantities is not a fixed number.

PROBLEM 2

Air at sea level and a dry-bulb temperature of 84°F has a relative humidity of 50%. The rate of heat needed to heat 2500 lbm/hr of this air to 102°F without changing the moisture content is most nearly

(A) 6500 Btu/hr

(B) 11,000 Btu/hr

(C) 23,000 Btu/hr

(D) 38,000 Btu/hr

Hint: Use a psychrometric chart to solve the problem.

PROBLEM 3

An unknown amount of propane (C_3H_8) is burned in the air. The theoretical stoichiometric reaction is

$$C_3H_8 + 5(O_2 + 3.773N_2) \rightarrow 4H_2O + 3CO_2 + 18.87N_2$$

A dry-product analysis shows that the reaction actually results in 6.6% O_2, 7.7% CO_2, and 2.2% CO. The analysis does not provide any information about water content. The actual reaction is found to be

$$3.3C_3H_8 + 22(O_2 + 3.773N_2)$$
$$\rightarrow 13.2H_2O + 7.7CO_2 + 6.6O_2 + 2.2CO + 83.5N_2$$

The excess air in the reaction is most nearly

(A) 14%

(B) 25%

(C) 33%

(D) 77%

Hint: Pay close attention to the number of moles of propane and air in each of the reactions.

PROBLEM 4

Consider two Brayton cycles. In both systems, there is isentropic compression in the compressor (1–2) and constant-pressure heat addition in the combustor (2–3).

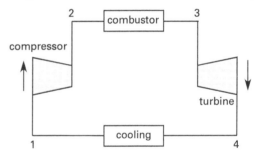

In system A, an isentropic expansion occurs in the turbine (3–4), followed by constant-pressure cooling (4–1). In system B, the turbine isentropic efficiency is 75%.

The two systems operate such that points 1, 2, and 3 are the same on a temperature-entropy (*T-s*) diagram for each. Which of the following statements are true?

I. The thermal efficiency of system A is higher than that of system B.

II. The work of the compressor is the same in systems A and B.

III. More heat is rejected during cooling in system A than in system B.

IV. The work out of the turbine is greater in system A than in system B.

(A) I and II only

(B) III and IV only

(C) I, II, and IV only

(D) I, II, III, and IV

Hint: No calculations are required.

PROBLEM 5

In Europe, the standard test for a ground source heat pump uses a high temperature of 95°F (35°C) and a low temperature of 32°F (0°C). The amount of energy that can be extracted for each kilowatt-hour of electricity used is most nearly

(A) 1.5 kW-hr

(B) 3.0 kW-hr

(C) 5.8 kW-hr

(D) 8.8 kW-hr

Hint: An ideal heat pump is a Carnot heat engine operating in reverse.

PROBLEM 6

A jet propulsion system using air as the fluid operates under the conditions shown.

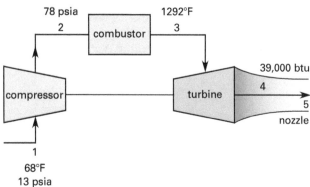

The efficiency of the compressor is 75%. Assume isentropic efficiency. The work input into the compressor per unit mass is most nearly

(A) 14 Btu/lbm

(B) 110 Btu/lbm

(C) 260 Btu/lbm

(D) 320 Btu/lbm

Hint: Begin by assuming the compressor is an isentropic process.

PROBLEM 7

A preliminary design to transport 1000 gpm of water from a reservoir to an open tank is shown. The line includes four shutoff gate valves, one swing check valve, and two 45° elbows (see *Illustration for Problem 7*).

The surface roughness of the pipe is 100. The total dynamic head of the pump is most nearly

(A) 120 ft

(B) 160 ft

(C) 200 ft

(D) 230 ft

Hint: The friction head loss for each section of pipe must be calculated separately.

PROBLEM 8

A Rankine cycle using a fluidized bed water boiler is being developed for use as a power plant. The temperatures in the boiler tubes are limited to 700°F, and the condenser operates at 5 psia. An ideal Rankine cycle is used, and the turbine exhaust is saturated vapor.

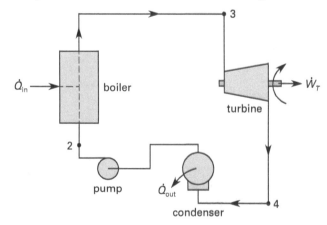

Illustration for Problem 7

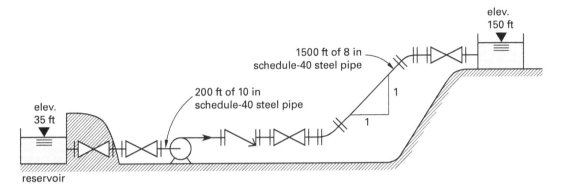

(not to scale)

The boiler pressure is most nearly

(A) 14 psia

(B) 40 psia

(C) 70 psia

(D) 200 psia

Hint: Only the steam tables are needed to solve the problem; no calculations are needed.

PROBLEM 9

The ideal turbojet shown operates at an altitude where the ambient pressure is 7.44 psi and the temperature is –4°F.

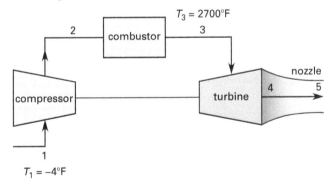

If the pressure ratio between states 1 and 2 is 8.0. The turbine outlet temperature, T_4, is most nearly

(A) 370°F

(B) 950°F

(C) 1900°F

(D) 2300°F

Hint: All the work of the turbine goes into driving the compressor.

PROBLEM 10

A one-pass, crossflow radiator uses water as the engine coolant and has the characteristics shown.

coefficient of heat transfer	62.9 W/m²·°C
air heat transfer surface area	3.1 m²

The radiator is being tested under the conditions shown.

water flow rate:	20 L/min
air velocity at the radiator face:	3 m/s
inlet air temperature	40°C
outlet air temperature	67°C
inlet water temperature	100°C
outlet water temperature	94°C

The heat transfer rate is most nearly

(A) 1.0 kW

(B) 8.3 kW

(C) 18 kW

(D) 37 kW

Hint: Use a correction factor of 1.0.

PROBLEM 11

The throttling area in a needle valve is a function of the

    I.   seat opening diameter

    II. needle lift

    III. cone height

    IV. cone half angle

    (A) I and IV

    (B) II and III

    (C) I, II, and IV

    (D) I, III, and IV

Hint: The throttling areas for a needle valve and a poppet valve are calculated using the same equation.

PROBLEM 12

A manufacturer of electrically actuated four-way hydraulic valves uses the following equation to compute the flow through its valves.

$$Q_R = KI_m\sqrt{p_v}$$

Q_R is the rated flow in gallons per minute, K is a valve constant, I_m is the maximum DC current supplied to the valve (15 mA), and p_v is the pressure drop across the valve. The manufacturer rates its valves assuming a 1000 psi drop across the valve. In a particular application, the supply and peak load pressures are 1500 psi and 1100 psi, respectively. The cost of the valve increases with size, and the valve must supply a flow rate of at least 23 in³/sec. The flow rate of the least expensive valve under these conditions is most nearly

    (A) 1 gal/min

    (B) 5 gal/min

    (C) 10 gal/min

    (D) 15 gal/min

Hint: The valve constant is key here.

PROBLEM 13

The supplementary firing system of a heat recovery steam generator burns natural gas in the flue gas stream with a thermal efficiency of 55%. The higher heating value of the natural gas used is 17,000 Btu/lbm. The mass flow rate of the flue gas through the firing system is 3,180,000 lbm/hr, and the flue gas enters the firing system at a temperature of 980°F. The combustion process in the firing system occurs at constant pressure,

and the specific heat of the flue gas before and after combustions is approximately 0.248 Btu/lbm-°F. The natural gas is fed into the firing system at a temperature of 60°F and a mass flow rate of 65,000 lbm/hr. The specific heat of the natural gas is 0.692 Btu/lbm-°F. The temperature of the flue gas as it leaves the system is most nearly

    (A) 700°F

    (B) 1700°F

    (C) 1800°F

    (D) 2100°F

Hint: Start by finding the thermal energy provided by the combustion of the natural gas and treating the HHV as an enthalpy term.

PROBLEM 14

The ideal vapor refrigeration cycle shown uses refrigerant-134a (R-134a) and operates between 20 psi and 120 psi.

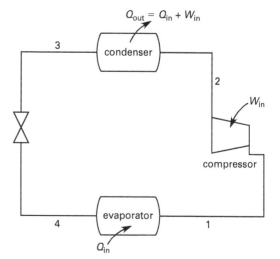

The cycle has the following properties.

state	p (psia)	h (Btu/lbm)	s (Btu/lbm-°R)
1 saturated vapor	20	102.73	0.22567
2 superheated vapor	120		0.22567
3 saturated liquid	120	41.787	
4 low-quality saturated mixture		41.787	

The system is to produce a cooling effect of 5 tons and operate with a coefficient of performance (COP) equal to 3.8. The compressor horsepower needed is most nearly

(A) 4.6 hp

(B) 6.2 hp

(C) 14 hp

(D) 30 hp

Hint: Use the COP equation to eliminate a table lookup.

PROBLEM 15

A jet taking off at sea level has its ideal turbojets operating as shown.

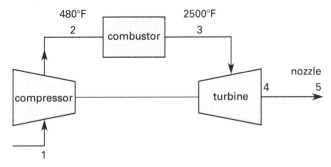

The mass flow rate is 20 lbm/sec. The rate of heat addition by the combustor is most nearly

(A) 10×10^3 Btu/hr

(B) 580×10^3 Btu/hr

(C) 35×10^6 Btu/hr

(D) 51×10^6 Btu/hr

Hint: This is an application of a Brayton Cycle.

PROBLEM 16

In a Rankine cycle with one open-feedwater heater, steam enters the high-pressure turbine at 1400 psia and 900°F. The steam expands through the high-pressure turbine to 100 psia, where 19.6% of the steam is extracted and diverted to an open-feedwater heater. The remaining steam expands through a low-pressure turbine to the condenser pressure of 2 psia. Saturated liquid leaves the open-feedwater heater and condenser (see *Illustration for Problem 16*).

The state values are shown.

state	p (psia)	T (°F)	h (Btu/ lbm)	s (Btu/lbm-° R)	x quality	y mass fraction
1	1400	900	1433.1	1.567	1.0	1.0
2	100		1159.1	1.567	0.968	0.196
3	100		1159.1	1.567	0.968	0.196
4	2	126.01	909.4	1.567	0.798	0.804
5	2	126.01	94.0	0.175	0.0	0.804
6	100	126.01	94.3	0.175	0.0	0.804
7	100		298.5	0.474	0.0	1.0
8	1400		302.8	0.474	0.0	1.0

The efficiency of the Rankine cycle is most nearly

(A) 28%

(B) 42%

(C) 58%

(D) 86%

Hint: Calculate the efficiency using values per pound-mass.

PROBLEM 17

An analysis of a turbine cycle shows that the heat transfer per unit mass is 340 Btu/lbm, the net work done per unit mass is 90 Btu/lbm, and the efficiency is 27%. An analysis of the heat exchanger shows that only 0.131 lbm of steam can be heated from 90°F to 930°F using 1 lbm of exhaust gases as the gases cool from 1535°R to 810°R. An analysis of the steam cycle shows that the net work done per unit mass is 570 Btu/lbm with an efficiency of 41%. The thermal efficiency of this combined cycle is most nearly

(A) 11%

(B) 22%

(C) 48%

(D) 83%

Hint: The ratio of the steam mass flow rate to the gas turbine mass flow rate will be needed.

Illustration for Problem 16

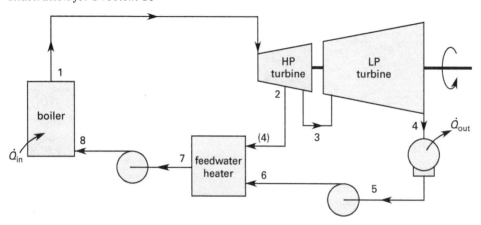

PROBLEM 18

A double-acting 3.0 in pneumatic cylinder with a 0.125 in diameter piston rod operates with a supply pressure of 100 psi. Assuming a typical efficiency of 0.85 to account for cylinder friction, the thrust generated on the return stroke is most nearly

(A) 200 lbf

(B) 300 lbf

(C) 600 lbf

(D) 700 lbf

Hint: Be sure to know the difference in geometry between the forward and return strokes.

PROBLEM 19

Water flows through a city's water supply system at a flow rate of 1750 gpm. The smallest diameter of schedule-40 steel pipe that should be used to transport the water through the system is most nearly

(A) 2.0 in

(B) 6.0 in

(C) 12 in

(D) 20 in

Hint: The recommended maximum velocity for water flowing through a city's water pipeline is 2 ft/sec to 5 ft/sec.

SOLUTION 1

To specify the state of a substance, an intensive property such as pressure, electrical potential, or magnetic potential is needed for each work mode, in addition to an intensive property to describe heat interactions.

The answer is (D).

Why Other Options Are Wrong

(A) This answer is incorrect because the number of independent quantities needed varies depending on the substance in the system.

(B) This answer is incorrect because the number of phases of the substance is irrelevant.

(C) This answer is incorrect because the number of phases of the substance is irrelevant.

SOLUTION 2

From a psychrometric chart, for 84°F air with 50% relative humidity, the enthalpy is 34.1 Btu/lbm, and the humidity ratio is 0.0125. [ASHRAE Psychrometric Chart No. 1 - Normal Temperature at Sea Level]

The moisture content is unchanged through the heating process, so the 102°F air has the same humidity ratio. From the psychrometric chart, the enthalpy for 102°F air with a 0.0125 humidity ratio is 38.6 Btu/lbm. [ASHRAE Psychrometric Chart No. 1 - Normal Temperature at Sea Level]

The heat needed is

$$q = \dot{m}\Delta h$$
$$= \left(2500 \ \frac{\text{lbm}}{\text{hr}}\right)\left(38.6 \ \frac{\text{Btu}}{\text{lbm}} - 34.1 \ \frac{\text{Btu}}{\text{lbm}}\right)$$
$$= 11{,}250 \ \text{Btu/hr} \quad (11{,}000 \ \text{Btu/hr})$$

The answer is (B).

Why Other Options Are Wrong

(A) This incorrect answer results when the 60% relative humidity line is used instead of the 50% relative humidity line.

(C) This incorrect answer results when a mass flow rate of 5200 lbm/hr is used instead of 2500.

(D) This incorrect answer results from incorrectly assuming that the relative humidity for both temperatures must be equal.

SOLUTION 3

The percent of excess air is given by

Combustion in Air

$$
\text{percent excess air} = \frac{(A/F)_{\text{actual}} - (A/F)_{\text{stoichiometric}}}{(A/F)_{\text{stoichiometric}}}
$$
$$
\times 100\%
$$

From the actual stoichiometric reaction, there are 22 moles of air for 3.3 moles of propane, and in the theoretical stoichiometric reaction, there are 5 moles of air for 1 mole of propane.

$$
\text{percent excess air} = \frac{\dfrac{22}{3.3} - \dfrac{5}{1}}{\dfrac{5}{1}} \times 100\%
$$
$$
= 0.3333 \quad (33\%)
$$

The answer is (C).

Why Other Options Are Wrong

(A) This incorrect answer results when the air-fuel ratios are added together and used in the denominator in the calculation of the percent excess air.

(B) This incorrect answer results when the excess air is calculated relative to the actual air instead of to the stoichiometric air.

(D) This incorrect answer results when the moles of propane and air are neglected in the calculation of the percent excess air.

SOLUTION 4

Plot both cycles on the same T-s diagram.

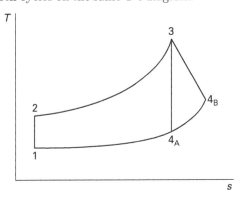

Thermal efficiency is defined as

Brayton Cycle (Steady-Flow Cycle)

$$
\eta = \frac{\dot{W}_{\text{net}}}{\dot{Q}_{\text{in}}}
$$
$$
= \frac{\dot{W}_{\text{turbine}} - \dot{W}_{\text{compressor}}}{\dot{Q}_{\text{in}}}
$$

$\dot{Q}_{2,\text{in}}$ depends on T_3 and T_2, and $\dot{W}_{2,\text{compressor}}$ depends on T_1 and T_2, so both $\dot{Q}_{2,\text{in}}$ and $\dot{W}_{2,\text{compressor}}$ are the same for both cycles; however, $\dot{W}_{2,\text{turbine}}$ is related to T_4-T_3 and is larger for system A.

Therefore, the thermal efficiency of system A is higher, and statement I is true.

As just described, the work of the two compressors is the same, so statement II is true.

The heat rejected during cooling varies in proportion to $T_4 - T_1$, which is larger for system B. Thus, statement III is false.

The work out of the turbine varies in proportion to $T_3 - T_4$, which is larger for system A. Thus, statement IV is true.

The answer is (C).

Why Other Options Are Wrong

(A) This choice is wrong because statement IV is also true.

(B) This choice is wrong because statements I and II are true, and III is false.

(D) This choice is wrong because statement III is false.

SOLUTION 5

To calculate the coefficient of performance (COP), absolute temperatures are needed. Converting the high and low temperatures to degrees Rankine gives

Temperature Conversions

$$°R = °F + 459.69°$$
$$T_H = 95°\,F + 459.69° = 554.69°R$$
$$T_L = 32°\,F + 459.69° = 491.69°R$$

The COP is calculated using the equation shown.

Basic Cycles

$$COP = \frac{Q_H}{W}$$

The equation for the upper limit of COP for a heat pump based on the reversed Carnot cycle is

Basic Cycles

$$COP_c = \frac{T_H}{T_H - T_L}$$

Equate these two expressions for COP, and solve for the amount of energy that can be extracted.

$$\frac{Q_H}{W} = \frac{T_H}{T_H - T_L}$$
$$Q_H = W\left(\frac{T_H}{T_H - T_L}\right)$$
$$= (1\ kW\text{-}hr)\left(\frac{554.69°R}{554.69°R - 491.69°R}\right)$$
$$= 8.8\ kW\text{-}hr$$

The answer is (D).

Why Other Options Are Wrong

(A) This incorrect answer results when the temperatures are not converted to Rankine.

(B) This incorrect answer results when 0° is used for T_L.

(C) This incorrect answer results when the numerator is converted to Rankine, but the two temperatures in the denominator are not.

SOLUTION 6

Find the absolute temperature of the air entering the compressor.

Temperature Conversions

$$°R = °F + 459.69°$$
$$T_i = 68°F + 459.69° = 527.69°R$$

Use the isentropic flow relationships to find the exit temperature from the compressor. For air, the ratio of specific heats is $k = 1.40$. [Thermal and Physical Properties of Ideal Gases (at Room Temperature)]

Isentropic Flow Relationships

$$\frac{p_2}{p_1} = \left(\frac{T_2}{T_1}\right)^{k/(k-1)}$$
$$T_2 = T_1\left(\frac{p_2}{p_1}\right)^{(k-1)/k}$$
$$= (527.69°R)\left(\frac{78\ \dfrac{lbf}{in^2}}{13\ \dfrac{lbf}{in^2}}\right)^{(1.40-1)/1.40}$$
$$= 881°R$$

Converting to degrees Fahrenheit,

$$T_2 = 881°R - 459.69° = 421.31°F$$

The work per unit mass done by the compressor is

Compressors

$$w_{comp} = -c_p(T_e - T_i)$$

Thus, the work input into the compressor operating at an efficiency η is

$$w_{comp} = \frac{-c_p(T_2 - T_1)}{\eta}$$

The specific heat of air at this temperature range is about 0.240 Btu/lbm-°F. [Thermal and Physical Properties of Ideal Gases (at Room Temperature)]

The work input is

$$w_{comp} = \frac{\left(0.240\ \dfrac{Btu}{lbm\text{-}°F}\right)(421.31°F - 68°F)}{0.75}$$
$$= 113\ Btu/lbm \quad (110\ Btu/lbm)$$

The answer is (B).

Why Other Options Are Wrong

(A) This incorrect answer results when T_1 is not converted to Rankine in the isentropic flow relationship.

(C) This incorrect answer results when 881°R is not converted to degrees Fahrenheit.

(D) This incorrect answer results when the ratio 78 psia/13 psia is raised to the 0.4 power.

SOLUTION 7

The pump head, h_{pump}, is equal to the difference in energy between the two sides of the pump. It can be calculated by modifying the Bernoulli equation to include this quantity and the friction head loss, h_f.

Bernoulli Equation

$$\frac{p}{\rho} + \frac{v^2}{2g_c} + \frac{gz}{g_c} = \text{constant}$$

$$\frac{p_1}{\rho} + \frac{v_1^2}{2g_c} + \frac{z_1 g}{g_c} + \frac{h_{pump} g}{g_c} = \frac{p_2}{\rho} + \frac{v_2^2}{2g_c} + \frac{z_2 g}{g_c} + \frac{h_f g}{g_c}$$

With the reservoir and tank surfaces as points 1 and 2, respectively, the pressures at both points are atmospheric, so the pressure terms are equal and cancel out. The velocities at both points are negligible, so the velocity terms both equal zero and cancel out. Simplifying the Bernoulli equation and solving for the pump head gives

$$h_{pump} = (z_2 - z_1) + h_f$$

The head loss due to friction, h_f, can be found using a steel pipe friction table. Since the suction and discharge lines are of different diameters, calculate the friction head loss for each section separately, and then add them together to find the total friction head loss. Because the pipe lengths are long and there are few fittings, the losses due to the fittings can be ignored in determining an approximation that can be used to select one of the answers.

From the steel pipe friction table, for a flow of 1000 gpm, the head loss of the 10 in diameter suction line is 1.0 ft per 100 ft, and the head loss of the 8 in diameter discharge line is 3.0 ft per 100 ft. [Steel Pipe Friction Tables]

The total head loss is

$$h_{f,\text{suction}} = \left(\frac{1.0 \text{ ft}}{100 \text{ ft}}\right)(200 \text{ ft}) = 2.0 \text{ ft}$$

$$h_{f,\text{discharge}} = \left(\frac{3.0 \text{ ft}}{100 \text{ ft}}\right)(1500 \text{ ft}) = 45 \text{ ft}$$

$$h_{f,\text{total}} = 2.0 \text{ ft} + 45 \text{ ft} = 47 \text{ ft}$$

The pump head is

$$\begin{aligned} h_{pump} &= (z_2 - z_1) + h_f \\ &= (150 \text{ ft} - 35 \text{ ft}) + 47 \text{ ft} \\ &= 162 \text{ ft} \quad (160 \text{ ft}) \end{aligned}$$

The answer is (B).

Why Other Options Are Wrong

(B) This incorrect answer results when the pipe friction is neglected.

(C) This incorrect answer results when the value of 35 ft from the left side of the Bernoulli equation is not subtracted from the right side of the equation.

(D) This incorrect answer results when the value of 35 ft from the left side of the Bernoulli equation is added to, not subtracted from, the right side of the equation.

SOLUTION 8

Saturated vapor exists at point 4. Use the steam table to find the entropy at point 4 (5 psia). [Properties of Saturated Water and Steam (Pressure) - I-P Units]

$$s_4 = 1.8440 \text{ Btu/lbm-°R}$$

Since the expansion inside a turbine is an isentropic process,

$$s_1 = s_4 = 1.8440 \text{ Btu/lbm-°R}$$

Using the superheated steam tables, an entropy value of 1.8440 Btu/lbm-°R at 700°F occurs at a pressure of 70.1 psia (70 psia). [Properties of Superheated Steam - I-P Units]

The answer is (C).

Why Other Options Are Wrong

(A) This incorrect answer results when the first value of 1.840 is encountered in the superheated steam tables.

(B) This incorrect answer results when the entropy value of 1.844 at 580°F is used.

(D) This incorrect answer results when the entropy at a saturation pressure of 20 psig is used.

SOLUTION 9

Find the exit temperature, $T_e = T_2$, by using the isentropic flow relationships. For air, the ratio of specific heats is $k = 1.40$. [Thermal and Physical Properties of Ideal Gases (at Room Temperature)]

Isentropic Flow Relationships

$$\frac{p_2}{p_1} = \left(\frac{T_2}{T_1}\right)^{\frac{k}{(k-1)}}$$

$$8.0 = \left(\frac{T_2}{-4°F + 460°}\right)^{\frac{1.40}{(1.40-1)}}$$

$$T_2 = 826°R$$

The absolute temperatures at the compressor inlet and the turbine inlet are

$$T_1 = -4°F + 459.69° = 455.69°R$$
$$T_3 = 2700°F + 459.69° = 3159.69°R$$

For the turbine, the work per unit mass is

Turbines

$$w_{turb} = c_p(T_i - T_e)$$

For the compressor, the work per unit mass is

Compressors

$$w_{comp} = -c_p(T_e - T_i)$$

All the work of the turbine goes into driving the compressor.

$$|w_{turb}| = |w_{comp}|$$
$$c_p(T_{i,turb} - T_{e,turb}) = c_p(T_{e,comp} - T_{i,comp})$$
$$c_p(T_3 - T_4) = c_p(T_2 - T_1)$$
$$T_4 = T_3 + T_1 - T_2$$
$$= 3159.69°R + 455.69°R - 826°R$$
$$= 2790°R$$

Converting back to degrees Fahrenheit gives

$$2790°R - 459.69° = 2330.31°F \quad (2300°F)$$

The answer is (D).

Why Other Options Are Wrong

(A) This incorrect answer results when the value of T_2 is assumed to be the turbine exit temperature.

(B) This incorrect answer results when, in the calculation of the exit temperature, 460° is subtracted from 2700°R and -4°F when calculating T_4.

(C) This incorrect answer results when the temperatures are not converted to Rankine when calculating T_4.

SOLUTION 10

The counterflow configuration is as shown.

Calculate the log mean temperature difference for the counterflow heat exchanger.

Log Mean Temperature Difference (LMTD)

$$\Delta T_{lm} = \frac{(T_{Ho} - T_{Ci}) - (T_{Hi} - T_{Co})}{\ln \frac{T_{Ho} - T_{Ci}}{T_{Hi} - T_{Co}}}$$

$$= \frac{(94°C - 40°C) - (100°C - 67°C)}{\ln \frac{94°C - 40°C}{100°C - 67°C}}$$

$$= 42.6°C$$

The heat transfer rate is

Rate of Heat Transfer

$$\dot{Q} = UAF\Delta T_{lm}$$
$$= \left(62.9 \ \frac{W}{m^2 \cdot °C}\right)(3.1 \ m^2)(1.0)(42.6°C)$$
$$= 8307 \ W \quad (8.3 \ kW)$$

The answer is (B).

Why Other Options Are Wrong

(A) This incorrect answer results when the area $0.37 \ m^2$ is used.

(C) This incorrect answer results when the log mean temperature difference is calculated incorrectly.

(D) This incorrect answer results when incorrect values are used for the coefficient of heat transfer and the log mean temperature difference.

SOLUTION 11

The throttling area in a needle valve is a function of the seat opening diameter, needle lift, and cone half angle.

The answer is (C).

Why Other Options Are Wrong

(A) This answer is incorrect because the throttling area is also a function of needle lift.

(B) This answer is incorrect because the throttling area is also a function of seat opening diameter, not a function of cone height.

(D) This answer is incorrect because the throttling area is also a function of needle lift, but not a function of cone height.

SOLUTION 12

Convert the needed flow rate of 23 in^3/sec to gallons per minute. [Measurement Relationships]

$$Q_R = \frac{\left(23 \ \dfrac{\text{in}^3}{\text{sec}}\right)\left(60 \ \dfrac{\text{sec}}{\text{min}}\right)\left(7.481 \ \dfrac{\text{gal}}{\text{ft}^3}\right)}{\left(12 \ \dfrac{\text{in}}{\text{ft}}\right)^3}$$

$$= 5.97 \ \text{gal/min}$$

Begin by testing the capacity of the 5 gal/min valve because it is one of the middle answer options. Rearranging the manufacturer's equation to find the product of the constant, K, and the maximum DC current, I_m, using the manufacturer's equation and pressure drop value gives

$$Q_R = KI_m\sqrt{p_v}$$

$$KI_m = \frac{Q_R}{\sqrt{p_v}} = \frac{5 \ \dfrac{\text{gal}}{\text{min}}}{\sqrt{1000 \ \dfrac{\text{lbf}}{\text{in}^2}}}$$

$$= 0.158 \ \text{gal-in/min-lbf}^{1/2}$$

Determine the flow rate through the 5 gal/min valve for the pressure drop.

$$Q_{5\,\text{gal/min}} = KI_m\sqrt{p_v}$$

$$= \left(0.158 \ \frac{\text{gal-in}}{\text{min-lbf}^{1/2}}\right)\left(\sqrt{1500 \ \frac{\text{lbf}}{\text{in}^2} - 1100 \ \frac{\text{lbf}}{\text{in}^2}}\right)$$

$$= 3.16 \ \text{gal/min}$$

The flow rate for the 5 gal/min valve is less than the needed flow rate of 5.97 gal/min, so the 5 gal/min valve is not large enough.

Check the capacity of the 10 gal/min valve. The constant for this valve is

$$KI_m = \frac{10 \ \dfrac{\text{gal}}{\text{min}}}{\sqrt{1000 \ \dfrac{\text{lbf}}{\text{in}^2}}} = 0.316 \ \text{gal-in/min-lbf}^{1/2}$$

Determine the flow rate through the 10 gal/min valve for the pressure drop.

$$Q_{10\,\text{gal/min}} = \left(0.316 \ \frac{\text{gal-in}}{\text{min-lbf}^{1/2}}\right)\left(\sqrt{1500 \ \frac{\text{lbf}}{\text{in}^2} - 1100 \ \frac{\text{lbf}}{\text{in}^2}}\right)$$

$$= 6.32 \ \text{gal/min}$$

This is greater than the needed flow rate, so the 10 gal/min valve will suffice. The 15 gal/min valve would also provide a large enough flow rate, but it is more expensive than the 10 gal/min valve.

The answer is (C).

Why Other Options Are Wrong

(A) This incorrect answer results from not converting seconds to minutes in the calculation of the needed flow rate.

(B) This incorrect answer results from overlooking that the flow rate of 5 gal/min is based on a pressure drop of 1000 psi. In this particular application, the pressure drop is only 400 psi, so the actual flow rate through the valve would be significantly less than 5 gal/min.

(D) This incorrect answer results from overlooking that this valve is not the least expensive one for the application.

SOLUTION 13

The thermal energy provided by the combustion of natural gas is

$$q_{\text{comb}} = \eta\dot{m}h_{\text{HHV}}$$

$$= (0.55)\left(65{,}000 \ \frac{\text{lbm}}{\text{hr}}\right)\left(17{,}000 \ \frac{\text{Btu}}{\text{lbm}}\right)$$

$$= 607{,}750{,}000 \ \text{Btu/hr}$$

The amount of thermal energy needed to increase the temperature of the natural gas from 60°F to 980°F is

$$q_{\text{temp}} = \dot{m}c_p(T_2 - T_1)$$

$$= \left(65{,}000 \ \frac{\text{lbm}}{\text{hr}}\right)\left(0.692 \ \frac{\text{Btu}}{\text{lbm-°F}}\right)$$

$$\times (980°\text{F} - 60°\text{F})$$

$$= 41{,}381{,}600 \ \text{Btu/hr}$$

The thermal energy available to heat the flue gas is

$$\Delta q = q_{\text{comb}} - q_{\text{temp}}$$

$$= 607{,}750{,}000 \ \frac{\text{Btu}}{\text{hr}} - 41{,}381{,}600 \ \frac{\text{Btu}}{\text{hr}}$$

$$= 566{,}368{,}400 \ \text{Btu/hr}$$

The mass flow rate of the flue gas after addition of the natural gas flow is

$$\dot{m} = \dot{m}_{\text{flue}} + \dot{m}_{\text{feed}}$$

$$= 3{,}180{,}000 \ \frac{\text{lbm}}{\text{hr}} + 65{,}000 \ \frac{\text{lbm}}{\text{hr}}$$

$$= 3{,}245{,}000 \ \text{lbm/hr}$$

The amount of energy needed to increase the flue gas temperature is given by the equation

$$q = \dot{m}c_p(T_{f2} - T_{f1})$$

Solving for T_{f2},

$$T_{f2} = \frac{q}{\dot{m}c_p} + T_{f1}$$

$$= \frac{566{,}368{,}400 \ \dfrac{\text{Btu}}{\text{hr}}}{\left(3{,}245{,}000 \ \dfrac{\text{lbm}}{\text{hr}}\right)\left(0.248 \ \dfrac{\text{Btu}}{\text{hr-°F}}\right)} + 980°\text{F}$$

$$= 1683°\text{F} \quad (1700°\text{F})$$

The answer is (B).

Why Other Options Are Wrong

(A) This incorrect answer results when the 980°F temperature increase is neglected.

(C) This incorrect answer results when the energy needed to heat the natural gas and the change in mass flow rate from the addition of the natural gas are neglected.

(D) This incorrect answer results when temperatures are mistakenly converted to Rankine.

SOLUTION 14

The enthalpy at state 1 is given in the problem statement as 102.73 Btu/lbm.

The enthalpy at state 2 can be calculated from the coefficient of performance equation.

Refrigeration Cycle - Single Stage

$$\text{COP}_{\text{ref}} = \frac{h_1 - h_4}{h_2 - h_1}$$

$$h_2 = \frac{h_1 - h_4}{\text{COP}_{\text{ref}}} + h_1$$

$$= \frac{102.73 \ \dfrac{\text{Btu}}{\text{lbm}} - 41.787 \ \dfrac{\text{Btu}}{\text{lbm}}}{3.8} + 102.73 \ \frac{\text{Btu}}{\text{lbm}}$$

$$= 118.8 \ \text{Btu/lbm}$$

Find the mass flow rate of refrigerant by rearranging the equation for the heat flow into the evaporator. [Measurement Relationships]

$$_4\dot{Q}_1 = \dot{m}(h_1 - h_4)$$

$$\dot{m} = \frac{_4\dot{Q}_1}{h_1 - h_4}$$

$$= \frac{(5 \ \text{tons})\left(12{,}000 \ \dfrac{\dfrac{\text{Btu}}{\text{hr}}}{\text{ton}}\right)}{102.73 \ \dfrac{\text{Btu}}{\text{lbm}} - 41.787 \ \dfrac{\text{Btu}}{\text{lbm}}}$$

$$= 984.5 \ \text{lbm/hr}$$

Calculate the compressor horsepower. [Measurement Relationships]

Compressors

$$P_{\text{comp}} = \dot{m}(h_2 - h_1)$$

$$= \frac{\left(984.5 \ \dfrac{\text{lbm}}{\text{hr}}\right)\left(118.8 \ \dfrac{\text{Btu}}{\text{lbm}} - 102.73 \ \dfrac{\text{Btu}}{\text{lbm}}\right)}{2545 \ \dfrac{\text{Btu}}{\text{hp-hr}}}$$

$$= 6.216 \ \text{hp} \quad (6.2 \ \text{hp})$$

The answer is (B).

Why Other Options Are Wrong

(A) This incorrect answer results when the answer is calculated in kilowatts instead of horsepower.

(C) This incorrect answer results when h_4 is added to h_1 in the numerator of the COP equation instead of subtracted.

(D) This incorrect answer results when the value for h_4 instead of h_1 is used in the final calculation.

SOLUTION 15

From a table of thermal and physical properties of ideal gases, the specific heat of air is 0.240 Btu/lbm-°R. [Thermal and Physical Properties of Ideal Gases (at Room Temperature)]

Find the rate of heat added by the combustor. Because the difference in temperatures is the same in degrees Fahrenheit or degrees Rankine, the temperatures do not need to be converted.

Brayton Cycle (Steady-Flow Cycle)

$$\dot{Q} = \dot{m}c_p(T_3 - T_2)$$
$$= \left(20 \ \frac{\text{lbm}}{\text{sec}}\right)\left(0.240 \ \frac{\text{Btu}}{\text{lbm-°R}}\right)$$
$$\times \left(2500°F - 480°F\right)\left(60 \ \frac{\text{sec}}{\text{min}}\right)\left(60 \ \frac{\text{min}}{\text{hr}}\right)$$
$$= 34.9 \times 10^6 \ \text{Btu / hr} \quad \left(35 \times 10^6 \ \text{Btu / hr}\right)$$

The answer is (C).

Why Other Options Are Wrong

(A) This incorrect answer results when Btu/sec is used as the solution.

(B) This incorrect answer results when the seconds-to-hours conversion is not squared.

(D) This incorrect answer results when 480°F is added to instead of subtracted from 2500°F.

SOLUTION 16

The efficiency of a Rankine cycle with regeneration is

Rankine Cycle With Regeneration

$$\eta = \frac{\dot{Q}_{\text{in}} - \dot{Q}_{\text{out}}}{\dot{Q}_{\text{in}}}$$
$$= \frac{(h_1 - h_8) - (1 - y_2)(h_4 - h_5)}{h_1 - h_8}$$
$$= \frac{\left(1433.1 \ \frac{\text{Btu}}{\text{lbm}} - 302.8 \ \frac{\text{Btu}}{\text{lbm}}\right) - (1 - 0.196)}{1433.1 \ \frac{\text{Btu}}{\text{lbm}} - 302.8 \ \frac{\text{Btu}}{\text{lbm}}}$$
$$\qquad \times \left(909.4 \ \frac{\text{Btu}}{\text{lbm}} - 94.0 \ \frac{\text{Btu}}{\text{lbm}}\right)$$
$$= 0.42 \quad (42\%)$$

The answer is (B).

Why Other Options Are Wrong

(A) This incorrect answer results when the mass fraction term $1 - y_2$ is neglected.

(C) This incorrect answer results when the term $h_1 - h_8$ is neglected.

(D) This incorrect answer results when only y_2, not $1 - y_2$, is used as the mass fraction term.

SOLUTION 17

The equation for the efficiency of the combined cycle is

Combined Cycle

$$\eta_c = \frac{\dot{W}_{\text{out}}}{\dot{Q}_{\text{in}}}$$

Efficiency can also be expressed in terms of work per unit mass.

$$\eta_c = \frac{w_{\text{out}}}{q_{\text{in}}}$$

The heat transfer into the system, q_{in}, is 340 Btu/lbm. Although the steam and gas mass flow rates are not given, their ratio is given as 0.131. Therefore, the net work done by the system per unit mass is

$$w_{\text{net}} = w_{\text{net,gas turbine}} + \left(\frac{m_{\text{steam}}}{m_{\text{gas}}}\right)w_{\text{net,steam cycle}}$$
$$= 90 \ \frac{\text{Btu}}{\text{lbm}} + (0.131)\left(570 \ \frac{\text{Btu}}{\text{lbm}}\right)$$
$$= 164.7 \ \text{Btu/lbm}$$

The efficiency of the combined cycle is

$$\eta_c = \frac{w_{\text{net}}}{q_{\text{in}}}$$
$$= \frac{164.7 \ \frac{\text{Btu}}{\text{lbm}}}{340 \ \frac{\text{Btu}}{\text{lbm}}}$$
$$= 0.4844 \quad (48\%)$$

The answer is (C).

Why Other Options Are Wrong

(A) This incorrect answer results from multiplying together the individual efficiencies of the gas turbine cycle and steam cycle.

Energy/Power System Apps.

(B) This incorrect answer results when only the net work per unit mass of the steam cycle, adjusted by the mass flow rate ratio, is used in the final calculation, instead of the net work done by the combined cycle.

(D) This incorrect answer results when the efficiency is calculated using the net work per unit mass of the steam cycle, instead of the heat transfer per unit mass.

SOLUTION 18

The equation for the thrust generated by the return stroke is

Force and Pressure to Retract Cylinder

$$F_L = \frac{\pi(D_1^2 - D_2^2)P_L}{4}$$

To account for efficiency, the previous equation must be modified.

$$F_L = \eta\frac{\pi(D_1^2 - D_2^2)P_L}{4}$$

$$= (0.85)\left[\frac{\pi\left[(3.0 \text{ in})^2 - (0.125 \text{ in})^2\right]\left(100\ \frac{\text{lbf}}{\text{in}^2}\right)}{4}\right]$$

$$= 599.8 \text{ lbf} \quad (600 \text{ lbf})$$

The answer is (C).

Why Other Options Are Wrong

(A) This incorrect answer results when the diameters are not squared in the calculation.

(B) This incorrect answer results when a pressure of 50 psi is inadvertently used instead of 100 psi.

(D) This incorrect answer results when an efficiency of 1.0 is used.

SOLUTION 19

From the continuity equation, the inner pipe diameter for the maximum recommended velocity is

Continuity Equation

$$Q = Av$$

$$\left(1750\ \frac{\text{gal}}{\text{min}}\right)\left(\frac{1 \text{ ft}^3}{7.48 \text{ gal}}\right)$$
$$\times\left(\frac{1 \text{ min}}{60 \text{ sec}}\right)\left(12\ \frac{\text{in}}{\text{ft}}\right)^3 = \left(\frac{\pi}{4}d^2\right)\left(5\ \frac{\text{ft}}{\text{sec}}\right)\left(12\ \frac{\text{in}}{\text{ft}}\right)$$

$$d = 11.9 \text{ in}$$

The smallest diameter of schedule-40 steel pipe with the required inner diameter is 12 in. [Pipe and Tube Data]

The answer is (C).

Why Other Options Are Wrong

(A) This incorrect answer results from using the conversion from hours to seconds instead of from minutes to seconds.

(B) This incorrect answer uses the pipe radius instead of the diameter.

(D) This incorrect answer results from using the minimum recommended velocity of 2 ft/sec instead of the maximum of 5 ft/sec.

Energy/Power System Apps.